P9-CFX-894

Other books by Peter Wohlleben

The Hidden Life of Trees:
What They Feel, How They Communicate—
Discoveries from a Secret World

The Inner Life of Animals:
Love, Grief, and Compassion—Surprising
Observations of a Hidden World

The Hidden Life of Trees:
The Illustrated Edition

The Secret Wisdom of Nature

PETER WOHLLEBEN

translation by JANE BILLINGHURST

THE Secret Wisdom of NATURE

Trees, Animals, and the Extraordinary Balance of All Living Things

Stories from Science and Observation

DAVID SUZUKI INSTITUTE

GREYSTONE BOOKS

Vancouver/Berkeley

Originally published in Germany in 2017 as *Das Geheime Netzwerk der Natur*

First published in English by Greystone Books in 2019

19 20 21 22 23   6 5 4 3 2

Greystone Books Ltd.
www.greystonebooks.com

David Suzuki Institute
www.davidsuzukiinstitute.org

Cataloguing data available from Library and Archives Canada
ISBN 978-1-77164-388-7 (cloth)
ISBN 978-1-77164-389-4 (epub)

Copyediting by Shirarose Wilensky
Jacket and interior design by Nayeli Jimenez
Jacket illustration by Briana Garelli
Printed and bound in Canada on ancient-forest-friendly paper by Friesens

Greystone Books gratefully acknowledges the Musqueam, Squamish, and Tsleil-Waututh peoples on whose land our office is located.

Greystone Books thanks the Canada Council for the Arts, the British Columbia Arts Council, the Province of British Columbia through the Book Publishing Tax Credit, and the Government of Canada for supporting our publishing activities.

Canada

Contents

Introduction

NATURE IS LIKE the mechanism in an enormous clock. Everything is neatly arranged and interconnected. Every entity has its place and its function. Take the wolf, for example. Under the order Carnivora, there is the suborder Caniformia, which includes the family Canidae and the subfamily Caninae, which includes the genus *Canis*, and within that genus is the species wolf. Phew. As predators, wolves regulate the number of plant eaters so that deer populations, for example, do not multiply too rapidly. All animals and plants are held in a delicate balance, and every entity has its purpose and role in its ecosystem. This way of organizing life supposedly gives us a clear view of the world, and thus a sense of security. As erstwhile plains dwellers, our most important sense is sight, and our species relies on viewing things clearly. But do we really have a clear view of what is going on?

The wolves remind me of a story from my childhood. I was about five years old and on vacation visiting my grandparents in Würzburg when my grandfather gave me an old clock. The first thing I did was take the clock apart, because I just couldn't wait to find out how it worked. Even though I was convinced that I knew how to put it back together in working order, I couldn't do it. After all, I was just a young child. After I rebuilt it, there were a few cogs left over—and a grandfather who was not in the best of moods. In the wild, wolves play the role of such cogs. If we eradicate them, not only do the enemies of sheep and cattle ranchers disappear, but the finely tuned mechanism of nature also begins to run differently, so differently that rivers change course and many local bird species die out.

Things can also go awry when a species is added; for example, when the introduction of a nonnative fish leads to a massive reduction in the local elk population. Because of a fish? The earth's ecosystems, it seems, are a bit too complex for us to compartmentalize them and draw up simple rules of cause and effect. Even conservation measures can have unexpected results. Who knew, for example, that recovering crane populations in Europe would affect the production of Iberian ham?

And so it's high time we took a good look at the interconnections between species both large and small. And if we do that, we get a chance to contemplate odd creatures, such as nocturnal red-headed flies that take wing only in winter on the lookout for old bones, or beetles that seek out cavities in rotting trees, where they dine on the feathery remains of pigeons and owls (but only when they're mixed together). The more light you shed on relationships between species, the more fascinating facts you reveal.

But nature is much more complex than a clock, isn't it? In nature, not only does one cog connect with another; everything is also connected by a network so intricate that we will probably never grasp it in its entirety. And that is a good thing, because it means that plants and animals will always amaze us. It's important for us to realize that even small interventions can have huge consequences, and we'd do better to keep our hands off everything in nature that we do not absolutely have to touch.

So you can get a clearer picture of this intricate network, I'd like to show you some examples. Let's be amazed together.

1

Of Wolves, Bears, and Fish

WOLVES ARE A wonderful example of how complex connections in nature can be. Amazingly enough, these predators are able to reshape riverbanks and change the course of rivers.

The matter of changing the course of rivers happened in Yellowstone, the very first national park in the United States. In the nineteenth century, people began the process of eradicating wolves in the park, primarily in response to pressure from ranchers in the surrounding area, who were worried about their grazing livestock. The last pack was wiped out in 1926. Individual wolves continued to be spotted occasionally until the 1930s, when they, too, were eliminated. Other animals living in the park were either spared or, in some cases, actively encouraged. In harsh winters, for example, rangers went so far as to feed the elk.

Changes came quickly. No sooner was pressure from predators lifted than elk populations began to increase steadily, and large areas of the park were stripped bare. Riverbanks were particularly hard hit. The juicy grass at the river's edge disappeared, along with all the saplings growing there. The desolate landscape didn't even provide enough food for birds, and the number of species drastically declined. Beavers joined the ranks of the losers, because beavers depend not only on water but also on the trees that grow close by. Willows and poplars are some of their favorite foods. They cut them down so that they can reach the trees' nutrient-rich new growth, which they then devour with great relish. Because all the young deciduous trees along the water's edge were ending up in the stomachs of hungry elk, the beavers had nothing to gnaw on—and they disappeared.

Riverbanks became wastelands, and because there was no longer any vegetation to protect the ground, seasonal flooding washed away ever-increasing quantities of soil—and erosion advanced rapidly. As a result, the rivers began to meander more and follow increasingly winding routes through the landscape. The less protection there is for the underlying layers of soil, the stronger the serpentine effect, especially on flat ground.

This sorry state of affairs continued for decades or, to be more precise, until 1995. This was the year wolves caught in Canada were released in Yellowstone to restore the park's ecological balance. What happened in the years that followed, and continues to this day, is what scientists call a trophic cascade. Basically, this means a change in the entire ecosystem via the food chain, starting at the top. The wolf was now at the top of the food chain, and what it triggered could perhaps better be described as a trophic avalanche.

The wolves did what we all do when we're hungry: they found something to eat. What they found in the park were large numbers of easy-to-catch elk. It seems clear where this story is headed: the wolves ate the elk and elk numbers declined drastically, which gave little trees a chance to grow again. Does that mean that the key to solving the problem of disappearing trees is to have wolves instead of elk? Thankfully, it's not nature's way to simply swap out one animal for another, and here's why. The fewer elk there are, the longer it takes the wolves to find them, and below a certain residual number, it's no longer worth the wolves' while to hunt elk, which means that the wolves must either leave the park or starve.

In Yellowstone, however, in addition to declining elk numbers, there was something else going on. Thanks to the presence of wolves, the elk's behavior was changing, and what was triggering this change was fear. Elk began avoiding open areas along the riverbanks, retreating to places that offered better cover. True, they did come to the water from time to time, but they no longer stayed long, and when they were there, they were constantly scanning the landscape, worried they might spot one of the gray-coated hunters. Constant surveillance left them little time to put their heads down among the willow and poplar saplings that were now growing abundantly along the riverbanks. Both trees are so-called pioneer species, and they grow faster than most: it's not unusual for them to grow 3 feet in a single season.

Within a few years, the riverbanks became stable once again. Stable banks slowed the flow of rivers, which, in turn, carried off less soil. The meandering stopped, though the serpentine curves the rivers had already carved into the landscape remained. Most importantly, the beavers' food sources returned, and the industrious little creatures began to build dams, which slowed the flow of water even more. Increasing numbers of ponds formed, which created small paradises for amphibians. In this blossoming of diversity, the number of bird species also increased substantially. (You can find an impressive video about this on the home page for Yellowstone National Park.)[1]

There are some, however, who question this interpretation. At the same time the wolves returned, a multiyear drought ended, and heavier rainfalls improved conditions for the trees—both willows and poplars love moist soil. But this explanation for increased tree growth ignores the beavers. In places where these buck-toothed engineers live, variations in precipitation have little effect on tree growth, at least not close to riverbanks. Beaver dams hold back rivers, saturating their banks, which makes it easier for trees to find water, even if it hasn't rained for months. And it is precisely this process that was set in motion once again by the wolves: fewer elk close to the riverbanks = more willows and poplars = more beavers. Is everything clear now?

Unfortunately, I have to disappoint you, because things now get even more complicated. Some researchers believe it is simply the number of elk that is the problem and not the way they behave. In this scenario, there are fewer elk in the park overall since the reintroduction of wolves (because so many of them have been eaten); therefore, you don't see so many of them near the rivers.

Are you completely confused by now? No wonder. I have to admit that for a while there even I felt like that five-year-old I mentioned in the introduction. In the case of Yellowstone, there's no doubt that because human intervention has been dialed back, the clock is slowly beginning to tick again. And if scientists don't yet understand the process down to the last detail, even that is an encouraging admission. The deeper the realization that even the smallest disturbance can lead to unpredictable changes, the stronger the arguments are in favor of protecting larger areas.

The reintroduction of wolves has done more than just help the trees and the inhabitants of the riverbanks. Other predators have benefited, as well. Things were not looking so good for grizzlies in the decades when the elk population was exploding. In the fall, grizzly bears depend on berries. They feed tirelessly on these tiny sugar- and carb-filled power snacks so that they can pack on the pounds before winter. At some point, however, the small shrubs with their apparently inexhaustible supplies no longer provided enough berries, or, rather, the bushes were being plundered by someone else—for elk also love the calorie-rich fruit. Now that wolves are hunting these large plant eaters again, there are more berries left for the bears when the time comes to harvest them in the fall. And since the arrival of the wolves, the bears have been in much better health.[2]

I began the wolves' story by saying that they were eradicated in response to pressure from ranchers. The wolves disappeared, but the ranchers did not. To this day, ranchers live around Yellowstone and graze their livestock on rangeland that runs right up to the park boundary. The attitude of many of them has not changed over the past decades either, so it's no surprise that wolves are shot the moment they leave the park. The number of wolves has dropped sharply in recent years, from a high point of 174 individuals in 2003 to about 100 individuals in 2016.

One reason for this reduction is improvements in technology. Many Yellowstone wolves now wear radio collars to help researchers locate the packs and find out how they move around the park—or when they leave it altogether. Elli Radinger, who has spent years observing wolves in Yellowstone, told me that people make illegal use of these same radio signals to shoot the animals the moment they leave the protection of the park. There's no more effective way of hunting down wolves, and it seems that hunters in Germany have also realized this. And so it was that in 2016, a young wolf wearing a radio collar was killed on the Lübtheener Heide nature reserve in Mecklenburg-Vorpommern.[3] Radio telemetry helps scientists get a better understanding of wolf movements, and it's ironic that the tool that is helping scientists understand wolves better so that they can protect them is also being used by hunters to track them down and kill them.

Despite the bad news, wolves are also ambassadors for optimism about conservation. It's almost unbelievable that wild animals this size can return to a region as densely populated as Central Europe and that the wolves are succeeding primarily because people living in the area not only accept their return but also hope for it. That is a blessing for all nature lovers and even more for nature itself.

In many areas, Central Europe finds itself in a similar situation to Yellowstone. Huge populations of deer and wild boar roam around as yet unconstrained by Wolf & Co. And, as was once the practice with elk in the park in America, the deer and boar in Europe are being fed large quantities of food. Harsh winters barely affect population levels, and even weak animals survive and happily procreate. The supplemental feeding programs are not courtesy of forest rangers, however. They are courtesy of hunters, who cart massive amounts of corn, beets, and hay into forests to ensure the open-air warehouse is always fully stocked with game they can hunt.

Rangers may not be trucking in food, but the forest service is playing its part. European forests are heavily exploited, and large-scale felling of trees allows so much light through to the forest floor that grasses and nonwoody plants spring up all over the place. This growth is effectively yet another supplemental feeding program that encourages further population growth. Today, the number of roe deer in German forests is at fifty times the level once found in the region's ancient forests, and red deer, originally animals of the plains, now hang out in the safety of the trees as people take over their ancestral ranges. The hosts of deer eat most of the saplings, meaning that in most places natural forest regeneration no longer happens.

Bad for the forest but good for the wolf. The returning wolves are finding a pantry stuffed full of tasty treats that have completely forgotten how to react appropriately to the new threat. For more than one hundred years, the human hunter has been their only enemy. People are slow and cannot hear very well compared to most forest dwellers. Sight is their keenest sense—during daylight, at least—and that is why countless generations of large plant eaters have learned that it's better to hide in the undergrowth by day and come out only at night. The tactic has worked so well that most people can hardly believe that, in relation to its size, Germany is home to more large wild mammals than almost any other country on Earth. And now along comes the wolf, bringing with it a completely different hunting technique.

The first thing wolves in Germany do is snap up "wimpy" prey such as mouflon sheep. Scientists argue about whether this sheep is truly wild or whether it is simply a domesticated sheep gone feral. Mouflon sheep were released centuries ago on islands in the Mediterranean, and they have now made their way to Central Europe. The reason for their advance is their impressive curved horns, which curl around to form an almost complete circle—a hunting trophy that looks good next to the deer antlers over the fireplace. Mouflon sheep are still being released into the wild today, even though this is illegal. (In most cases, the fence around their enclosure has been "breached.")

Whatever their history, mouflon sheep are not native to Central Europe, and recent developments support the contention that they might indeed be descended from domesticated stock: wherever wolves appear, the sheep disappear—into the wolves' stomachs. The sheep, it seems, have forgotten how to take evasive action. Another strike against them is that they are adapted to life in the mountains. Mouflon sheep are skilled climbers that evade their pursuers by taking refuge on steep cliffs where flatlanders like wolves do not stand a chance. In forests on level ground, the sheep cannot exploit this advantage, and when it comes to speed, they are hopelessly outmatched by the wolves. And so the mouflon sheep in Germany are caught unprepared and the natural order is reestablished.

Roe deer and red deer are next up. This probably surprises you. Not domesticated animals? If mouflon sheep are such easy prey, then what about domesticated sheep or goats or calves? After all, most of them are just standing around behind flimsy fences that might stop them from running away but are easy for wolves to crawl under or jump over. Rather than turning to sensationalized sources such as tabloid headlines, which just love to report alleged wolf attacks (more on that later), we would do better to peek over the shoulders of scientists. They are analyzing scat from Lusatian wolves in eastern Germany—one of the densest and longest established populations of these gray-coated hunters in the country.

After collecting thousands of samples, researchers at the Senckenberg Museum of Natural History in Görlitz came to the following conclusion: not sheep or goats but roe deer, at over 50 percent of the total mass, make up the lion's share of the wolves' diet. Red deer and wild boar account for about 40 percent, and, no, domesticated animals are still not next in line. That honor goes to hares and similar small mammals at around 4 percent. The fallow deer, which weighs in at 2 percent, is—much like the mouflon—an exotic species that has been released into the wild so that people can hunt it, and wolves enjoy sending them to the great hunting ground in the sky. Only now do we arrive at a few scattered livestock animals in the palette of prey, and they account for a meager 0.75 percent.[4]

You get a completely different impression if you leaf through the pages of the tabloid press. Here, reports of attacks on livestock dominate, and each and every one is worth a headline. Even before the results of genetic analysis have been published to ascertain whether the culprit really is a wolf and not perhaps a feral dog, the news is spread far and wide. If it turns out that the attacker was not a wolf after all, the correction usually appears only as a marginal note, and the public gets the impression that every goat and every sheep is henceforth in deadly danger.

But things don't need to be this way. It's relatively easy to keep wolves away from treasured livestock. In most cases, a simple electric fence is all it takes, and lots of farmers use them to contain their animals anyway. This kind of fence looks like a net with coarse mesh. The thin strands of metal woven into the netting are electrified when the fence is plugged into a charger.

We've enclosed our goat pasture at home with just such a fence, and many's the time I've forgotten to turn off the current when I entered the field. Ouch. The shock makes you feel as though you've been hit in the back with a plank. For days following such a misadventure, I overcompensate by checking over and over again to make sure there's no juice running through the wire before I touch it. It feels far worse for wolves, because they hit the barrier with their nose or ears. From then on, they prefer to help themselves to a side of venison or wild boar rather than expose themselves to such pain. The important thing is to make sure the fence is tall enough and in good working order. Some experts think 3 feet is sufficient, but my wife and I prefer to play it safe, and we opted for a 4-foot fence.

Elli Radinger, "my" wolf expert, told me that packs can change their preferred range of prey if older animals are shot and killed. Instead of hunting wild boar, roe deer, or red deer as they used to, they might turn to sheep and other domesticated animals. Wolf haters who want to keep wolves from attacking cattle would do better to leave their guns safely stashed in their cabinets.

There is yet one more thing wolves do: they bring a certain intensity to every forest experience. I remember clearly how happy and excited I was when I found wolf tracks one day. No, not here in Hümmel, where I live with my family, but on an isolated path through the forest in central Sweden. Just these tracks alone were enough to turn my forest walk into an adventure, and they made the forest itself seem a little bit wilder. And this is exactly the sensation I want to share with as many people as possible: the wolf restores the forest's wild soul. It is a sign that, even in heavily settled parts of the world, it's possible to allow the return of sizeable animals that disappeared long ago. In contrast to the situation in Yellowstone, wolves in Germany are returning of their own accord, moving in from Poland and slowly extending their range into one state after another.

Does that mean that you need to be afraid every time you go for a walk in the woods? Newspaper reports of troublesome wolves are stacking up. It's not that they've hurt anyone, but their presence close to villages or, worse yet, elementary schools is enough to make some people's blood run cold. It goes without saying that wolves are wild animals and it's not a good idea to pet them or cuddle them, but as long as we don't intentionally habituate them to our presence, the risks are small.

Unfortunately, there are always some people who are tempted to feed wolves. That is probably what happened with the wolves Kurti and Pumpak, which kept approaching villages close to Münster and in Lusatia respectively. Both wolves ended up being shot, even though they did nothing dangerous. In these cases, at fault were not the animals but the people who had been offering them food.

But let's consider the situation from a different perspective. How dangerous would it be if at some time in the future we had not a few hundred but a few thousand wolves loping through our forests? Strictly speaking, we've been living with a far more worrying situation for a long time now, because the countryside and our urban centers are already overrun by wolves. I'm talking about dogs, which are different from their ancestors in one important respect: they are no longer afraid of us. If you meet a wolf, it's probably just curious and will disappear when it finds out what it's dealing with. Wolves just don't think of us as prey.

And so it should come as no surprise when I tell you that of the two, dogs are the ones you should be less happy to encounter. If I had the choice between coming across a stray German shepherd or a wolf, I would opt for the wild animal. According to Olaf Tschimpke, president of the German conservation organization NABU, ten thousand incidents of dogs biting people are reported annually, and a few of these incidents are so severe that the victims die of their injuries.[5] Imagine, if you will, if just a fraction of these bites were inflicted by wolves. Someone would be demanding that all wolves be shot.

At the moment, however, it is not wolves but wild boar that are dominating the headlines in Germany. For example, in the middle of Berlin, the pigs are calmly plowing up lawns while homeowners just a few yards away anxiously try to drive them off by clapping loudly and shouting at them. Tulip beds laid waste, vineyards or cornfields stripped bare—all over the place, boar are responsible for economic losses and frustration. For years now, the boar population has been headed in just one direction: sharply up. Wild boar have no natural enemies here. Or, to be more precise, they had no natural enemies, because it is only recently that the wolf has reappeared as an adversary to be reckoned with.

One day, years ago, I was walking in a former opencut coal mining pit when I came across evidence of wolves. The pile of white bones and thick black hair was clearly the remains of a wild boar. I realized for the first time just how hard a wolf's life must be, and how it has to risk its life every time it needs to eat.

The pile of pig remains made me think of the drive hunts I used to participate in as a beater. One time, the dogs came across wild boar in the undergrowth and immediately gave chase. That evening, only three of the five dogs returned. The other two had probably lost their lives fighting the pigs. Many handlers that deploy hunting dogs insist that the local vet is informed and available. At the end of the day, after the work is done, some quickly sew their dogs' wounds themselves—wounds inflicted by the sharp tusks of wild boar.

For wolves, of course, even minor wounds can be life threatening, because a wolf that has to hunt with any kind of disadvantage is likely to starve. It is really admirable how these gray-coated hunters overcome all the dangers they encounter day in and day out over the course of their decade-long lives.

Before we leave the topic of wolves, I want to return once again to Yellowstone. "Yellowstone again?" I hear you ask. "Really?" Well, it could be any other random place in the world covered with vegetation and containing a wealth of animals. Central Europe would definitely fit the bill here. However, there has to be a large enough area—in this case, many thousands of square miles—where people are no longer manipulating the landscape in any way. Unfortunately, no such areas exist in Central Europe.

But what about national parks here? Aren't there plenty of areas being given this designation? True enough, but these preserves are tiny by nature's standards. In most of these protected spaces, there's not enough room for even one wolf pack, which means that there's no opportunity to observe natural processes at work. And even in national parks, unfortunately, huge interventions happen. For example, the largest clear-cuts in Germany—clear-cuts that are considerably larger than is normal in commercial forests—are carried out in some of these parks. The powers that be call these development zones.

Even if these clear-cuts are made with the best of intentions, it still means that people are constantly interfering with natural processes.

The only way to see how nature might surprise us is to simply sit back and let things take their course—perhaps while also carefully helping eradicated species reclaim territory and encouraging alien, introduced species to relinquish it. As this isn't happening in densely populated places such as Germany, we must look for these success stories in other parts of the world, which brings us back to Yellowstone.

This time, fish are the focal point of the story, or, to be more precise, lake trout. Lake trout are native to both Canada and the US (for example, in the Great Lakes), where their populations have declined significantly. As they are now endangered, there are costly hatchery programs to help sustain wild populations. However, these lake dwellers are not endangered everywhere in their range, and in some places, they are even a threat. No one knows who was responsible—whether it was anglers who wanted to increase the species available for them to fish or people who had a hazy grasp of how conservation works—but in the 1980s, lake trout suddenly appeared in Yellowstone Lake.

This would not be a problem if not for the fact that this ecosystem was already home to one of the lake trout's smaller relatives: the cutthroat trout. The name "cutthroat" comes from the fish's blood-red lower jaw, but it also describes the struggle they are now engaged in. The newcomers are outcompeting the fish that originally laid claim to this territory and crowding them out—and the struggle is affecting more than just the cutthroat. Amazingly enough, the elk in the park are also suffering because of this cutthroat competition.

So what is the connection between elk—which are strict vegetarians—and fish? Once again, an intermediary is the key to the puzzle. In this case, the grizzly bear. Grizzly bears love to eat cutthroat trout, which have become scarce since the introduction of lake trout. Cutthroat spawn in small streams, where they are easy to catch. The invasive trout behave quite differently. They shun crystal-clear tributaries and lay their eggs on the lake bed instead, which means that the exhausted parent fish end up well beyond the reach of grizzly bears. The bruins have to find other prey to fill their rumbling stomachs. This prey is trickier to hunt and is waiting for them on land. The bears now target elk calves, more and more of which meet their end after a blow from a well-armored grizzly bear paw. This is now happening so often that the elk population is declining noticeably.[6]

This is something to be celebrated, right? Weren't we just welcoming the return of the wolves for doing just this—reducing exploding elk populations? Aren't the bears doing the same thing in their own way? But, once again, the situation isn't quite that simple. Whereas the wolves hunt older animals, the bears target young ones, which drastically alters the age distribution of the herds. To put it another way: because of the bears, the elk population is aging, which speeds up the rate of population decline. Good for the trees, bad for the elk.

This case is another clear example that ecosystems are multifaceted and changes never affect just one species. Is it possible that the wolf doesn't have the greatest influence in

Yellowstone after all, and the prize should go to the trout-bear duo instead? That enormous clock, it seems, contains more cogs than we ever suspected.

Talking of fish: they interlock with the cogs in the forest in such an important way that they deserve a chapter all to themselves.

2

Salmon in the Trees

THE RELATIONSHIP BETWEEN trees and fish shows just how complicated ecosystems can be. Tree growth can be almost completely dependent on these flashes of silver, especially in areas where the soil is low in nutrients. Fish and rivers, it turns out, play an important role in nutrient distribution.

Let's take a look at salmon. Young salmon swim out into the ocean, where they remain for two to four years. They hunt and hang out, but mostly what they are doing is getting bigger and fatter. On the northwest coast of North America, there are a number of different species of salmon, of which the king salmon (also known as Chinook) is the largest. After its youthful years at sea, a full-grown king can measure up to 5 feet long and weigh up to 65 pounds. Not only has it built up a lot of muscle after scouring the vastness of the ocean in search of food, but it has also stored a lot of fat, which it will need to survive its arduous journey back to the stream where it was born.

Salmon battle their way against the current toward the headwaters of their natal rivers, sometimes for many hundreds of miles and up numerous waterfalls. They carry considerable quantities of nitrogen and phosphorus in their bodies, but these nutrients are of no interest to the salmon. They are toiling their way upstream so that they can spawn in the one and only frenzy of passion they will ever experience, and then finally breathe their last.

Over the course of their journey, the salmon's silvery skin loses its metallic sheen and takes on a reddish hue. The fish are no longer eating, and they are losing weight as they steadily deplete their stores of fat. Using the last of their strength, they mate in their natal streams, and then, exhausted, they die. For the forest and its inhabitants, the salmon runs mean it's time to get out and haul in the catch. Lining the riverbanks are hungry hunters—bears. Along the Pacific coast of North America, this means black bears and brown bears. The fish they catch from the rapids as the salmon fight their way upstream help them put on a thick layer of fat for the winter.

Depending on location and timing, the salmon have already lost some weight by the time they're caught. At first, the bears eat most of their catch, but later in the season, they get choosier. They still scoop skinny salmon out of the water—fish that have used up their reserves and therefore contain fewer calories—but if the fish don't contain much fat, the bears don't eat much of them, and the carcasses they discard give many other animals the opportunity to eat. Minks, foxes, birds of prey, and a myriad of insects pounce on the lightly nibbled remains and scatter them farther into the bush.

After mealtime, some parts of the salmon (such as the bones and the head) are left lying around to fertilize the soil directly. A lot of nitrogen is also distributed through the feces the animals expel after their feast. The amount of nitrogen that ends up in the forests alongside salmon streams is enormous. Scientists Scott M. Gende and Thomas P. Quinn reported that according to their detailed molecular analyses, up to 70 percent of the nitrogen in vegetation growing alongside the streams comes from the ocean—in other words, from salmon. Their data show that nitrogen from salmon speeds up the growth of trees so much that Sitka spruce in these areas grow up to three times faster than they would have without the fish fertilizer.[1] In some trees, more than 80 percent of the nitrogen they contain can be traced back to fish. How can we know this so precisely? The key is the isotope nitrogen-15, which, in the Pacific Northwest, is found almost exclusively in the ocean—or in fish that have spent time there. Therefore, finding evidence of these molecules in plants allows researchers to make a direct connection back to the source of the nitrogen—in this case, back to the salmon.

It turns out that not all the coveted nutrients remain on land. Eventually, they get eaten and digested, droppings containing the nutrients land on the ground, and the nutrients then gradually seep down into the soil. The trees are ready and waiting, and they eagerly suck them up with their roots. The trees are aided by fungi, which envelop the trees' feeder roots in a fine cottony web and help the trees take up a wide range of nutrients from the soil. Eventually, the trees shed their leaves or needles, and when these ancient giants die, their trunks rot into the ground. After an armada of organisms does a tidy job of breaking down the vegetative remains, the nutrients then enter the next tree whose turn it is to draw up the released elixirs of life from the soil. Not all of the nutrients remain trapped in the fine mesh of this web, however. Inevitably, some of them are washed down into the rivers and swept back out to the ocean, where innumerable tiny life-forms lie in wait for the shipment of food.

A striking study from Japan shows how important the trees' legacy is for the oceans. Katsuhiko Matsunaga, a marine chemist at Hokkaido University, discovered that fallen leaves leach acids into streams and rivers that are then swept down to the ocean. There, the acids fuel the growth of plankton, which are the first and most important link in the food chain. More fish because of the forest? The researcher recommended that local fishing companies plant trees along the coastline and riverbanks. More trees meant that more leaves fell into the water, and in time, increased tree cover led to increased numbers of fish and oysters for the local fishing companies to harvest.[2]

But let's get back to the salmon that fertilize Sitka spruce and other species that live in the forests of the Pacific Northwest. It's not only the trees that are indirect beneficiaries. For example, Dr. Tom Reimchen at the University of Victoria discovered that up to 50 percent of the nitrogen in some insects comes from fish.[3] The abundance of nutrients along salmon streams can be seen in the increased biodiversity of animals, plants, and birds, and the salmon scavengers (foxes, birds, and insects) become prey for other animals in the forest.

Dr. Reimchen and members of his team also took core samples from ancient trees. Their growth rings are like a historical archive: they reflect everything the tree has experienced over its lifetime. There are narrow rings for years of drought and correspondingly wider ones for years of ample rainfall, and of course, you can also work out the level of nutrients available to the tree. And so there's a direct connection between the number of fish in earlier times and the amount of that special isotope nitrogen-15 found in wood—and that's how core samples give us information about how many salmon once swam in these streams. It turns out that the number of salmon has declined dramatically over the last one hundred years, and many rivers in North America today have no salmon left in them at all.

But what does this story have to do with European forests? Quite a lot, if you're looking at how nature used to be. European rivers were also once full of salmon, and brown bears used to roam here, as well. Unfortunately, we can't test trees from those times for nitrogen from fish, because those trees are all gone now. From the Middle Ages on, the forests were either cut down or so heavily exploited that all the ancient trees have disappeared. The average age of beeches, oaks, spruce, or pines growing in Germany today is less than eighty. Eighty years ago, there were neither bears nor any salmon runs to speak of, so the wood in trees in Germany doesn't contain much nitrogen-15. But what about trees from earlier times? One way to find out how much nitrogen-15 they contained would be to test the beams in old timber-framed cottages, but, so far as I know, no one has done this.

There's no question, however, that salmon were once plentiful in Germany, and there's anecdotal evidence, such as a report that it was illegal to serve servants salmon more than three times a week.[4] The Atlantic salmon is the species native to Europe, and it is now returning to many different rivers thanks to the efforts of conservation organizations—particularly their efforts to clean up the waterways. I grew up close to the Rhine, and I remember that my parents didn't allow me to play in the water. Back then, chemical plants spewed out a cocktail of waste; the mix was so filthy that only a few species of fish survived.

Gradually, starting in the 1980s, regulations governing water quality were put in place. Even so, the federal minister for conservation, Klaus Töpfer, caused quite a stir in 1988, when he jumped into the Rhine to swim across it. Three years earlier, he had bet that thanks to new environmental policies, water quality would improve so much that the river would be fit for swimming again. The German weekly news magazine *Der Spiegel* reported rather derisively that the minister's eyes were bloodshot when he climbed out of the brown water. Apparently, the river wasn't quite as clean as he'd hoped it would be.[5]

Luckily, that has since changed. Now, as I write this, the Rhine is so clean that swimming beaches are appearing along its banks again. And salmon, too, are feeling at home in its waters, though they do still need help—quite a lot of help, actually. Adult salmon always swim back to the rivers of their youth, and when the salmon in a body of water die out, salmon almost never return there, because all the mature fish were born elsewhere.

To fix this, enterprising organizations release hundreds of thousands of young salmon into rivers where they can survive. However, it's not always easy to find such rivers, because dams and power plants impede the fish's progress in most of them. Many a turbine turns expensive hatchery-raised fish into sushi the moment they start their journey to the sea. For the return journey, there are fish ladders at dams, where the water splashes from rung to rung—that is to say, from pool to pool—imitating rapids the fish can climb by jumping up one pool at a time.

In the forest I manage, a lot of money was spent to make a small stream salmon friendly. The stream, which is only about 12 feet wide, had been sealed off by weirs for a long time. Its name—Armuthsbach or "stream of poverty"—is a testament to the straitened circumstances of bygone generations. Harnessing the power of the diverted water made it easier for people to grind what grain they had, and they also used the stream to fill ponds, where they raised fish. Eventually, the Armuthsbach ran out of water and dried up.

Salmon are just one example of the many water-loving species all the way down to crayfish and smaller freshwater crustaceans that can no longer move freely when dams get in their way. And if fish and other aquatic creatures can move only downstream but not upstream, then sooner or later there will be no large life-forms in the water above the dams. Dams are now being removed gradually so that fish can once again reach their spawning grounds. This is a huge accomplishment, and one that gives cause for hope. Indeed, adult salmon are constantly being spotted returning to the places where they were released so that they can spawn there after spending years out at sea. Finally, after a long absence, we will be getting the first generations of truly wild salmon born in rivers to which they will return.

THE SALMON ARE returning, but unfortunately, the bears are not. Bears would certainly be a problem in large urban centers along the Rhine, but it's not too much of a stretch to imagine them in rural areas. However, it doesn't have to be bears that distribute fish into the landscape. What about fish-eating birds such as cormorants? Cormorants were almost wiped out, but they are now coming back to rivers in Central Europe thanks to strong legal protection. I've been seeing them regularly on the Rhine and the Ahr since the 1990s. (The Ahr is a small tributary of the Rhine that has its source near the village of Hümmel, where I live, and the Armuthsbach is one of the streams that flows into it.)

Cormorants are skilled divers and excellent hunters underwater. After they've eaten their fill, they doze contentedly in the tops of the trees along the riverbank. As they doze, they defecate from time to time, and their droppings naturally contain valuable nitrogen. Of course, the value of their droppings depends on the number of birds, and too many perching at the same time can harm the trees. This is what happened in the Saarschleife, a dramatic hairpin turn in the Saar River, where people have created a North American–style forest near the riverbank by planting Douglas firs. (Douglas firs come from the Pacific coast of North America.) There's a whole colony of cormorants here, and the abundant quantity of excrement they discharge is so caustic that parts of the forest canopy are dying off, much to the dismay of local forest owners.

But that's not the main reason the birds have become so unpopular. The few salmon that battle their way upstream—products of costly reintroduction efforts—are often picked off by cormorants before they reach their spawning areas. So what happens next? A natural nutrient cycle ensues, but, as is so often the case, this cycle collides with the interests of the people who live here. I can understand that no one wants to stand idly by as cormorants threaten to destroy all the conservationists' hard work, but is that a good enough reason to automatically reach for a gun?

That is exactly what people did at the aforementioned Ahr, cheered on by members of the Ahr fishing syndicate, the organization that had worked so hard on behalf of the salmon. Is this perhaps a case of conservation clashing with nature? On its home page, the syndicate (a glance at the bylaws indicates that membership is restricted to anglers and those who own or rent fishing waters) specifically mentions that hunting the birds, which are strictly protected under the laws of the European Union, is still possible, thanks to an exemption intended to protect the fishing industry from financial losses.[6] It's a shame that this stand on cormorants taints the work of the organization, whose efforts on behalf of salmon are really commendable.

DO FORESTS THAT grow around developed areas in the world (and that includes practically all forests in Central Europe) even need natural fertilization with nitrogen? In recent decades, trees have completely new (and completely unnatural) sources of nitrogen available to them—a veritable deluge of them. In contrast to the clean air in the northern United States and Canada, the air in Central Europe is basically a murky soup. Perhaps not optically speaking, but it is in terms of pollutants. Or, I could say, in terms of "nutrients." Exhaust from vehicles and manure from agriculture provide more of these than plants want. But more of that later.

Nitrogen is naturally abundant in the air. You're breathing large quantities of it in and out as you read these lines. Oxygen, which is so vitally important for us, makes up just 21 percent. In contrast, nitrogen makes up 78 percent. Strictly speaking, three-quarters of every breath we take is useless—if we could separate out the gas we don't need. That doesn't mean that nitrogen doesn't matter to us. On the contrary, you're carrying around about 4.5 pounds of it in your body, processed into protein, amino acids, and other substances.[7]

It's pretty much the same with plants. Like us, they don't need nitrogen to breathe. What interests them are the special compounds in which it is held. These compounds are reactive, and they can be broken down and converted into protein or built into plants' genetic material. Unfortunately, these compounds tend to be rare in the natural world. If a tree doesn't have the good fortune to be growing alongside a salmon stream, it has a problem. Feces left by passing animals or perhaps even a whole carcass rotting within reach of its roots are causes for celebration.

Lightning plays its part by using its energy to combine atmospheric nitrogen with oxygen to create compounds plants can break down and absorb, and some trees and other plants have developed the ability to transform atmospheric nitrogen into plant-available compounds with the help of bacteria that live in special nodules on their roots. Alders, for example, manufacture their own fertilizer this way. Most species of trees, however, cannot do this and therefore rely on waste products from animals for their nitrogen needs.

Overall, it seems, nature regards these valuable nitrogen compounds as rare treats. Then we came along. Our modern internal combustion engines, in vehicles or heating systems, do the same thing lightning does. As a by-product of burning fossil fuels, they combine atmospheric nitrogen with oxygen to create compounds that travel long distances on the wind all over the world and are then washed back down to earth when it rains. Then there's agriculture, which forces the soil to be as productive as possible by adding synthetic fertilizers containing nitrogen. The quantity of nitrogen compounds released by human activity is significant. Every year, about 220 million tons worldwide rain down on the earth—about 60 pounds per person in the world and about 220 pounds per person in industrialized countries.[8]

Maybe that doesn't sound like much to you? Let's get back to the salmon and their beneficial effects on trees. A male chum salmon (also known as dog salmon or keta salmon) contains, on average, 4.5 ounces of nitrogen.[9] If Europeans were to calculate their nitrogen emissions in salmon, that would be about 750 fish per person per year. At 600 inhabitants per square mile—the population density in Germany—that would come to 450,000 salmon per square mile, and it's clear that such an enormous quantity of fish would completely overload the natural cycle. Exhaust fumes and applications of liquid manure and fertilizer reach the same threshold; however, for the most part, they are out of sight and therefore out of mind. They only appear on our radar when, one day, high levels of nitrogen compounds are detected in our drinking water.

Trees, however, have been aware of these emissions for a long time, and so have foresters. For decades, saplings have been growing markedly more quickly. This means that forests have been producing more timber, and forest production estimates now need new baselines. Foresters' yield charts—spreadsheets that indicate how quickly and at what age different tree species grow—have already had to be adjusted upward by 30 percent.

Is that a good sign? No, it's not. Left to their own devices, trees do not grow quickly. In undisturbed ancient forests, youngsters have to spend their first two hundred years waiting patiently in their mothers' shade. As they struggle to put on a few feet, they develop wood that is incredibly dense. In modern managed forests today, seedlings grow without any parental shade to slow them down. They shoot up and form large growth rings even without a nutrient boost from added nitrogen. Consequently, their woody cells are much larger than normal and contain much more air, which makes them susceptible to fungi—after all, fungi like to breathe, too. A tree that grows quickly rots quickly and therefore never has a chance to grow old. This process is now accelerating rapidly because of the extra nutrients in the air. The trees are like extreme athletes who are already doped up on steroids and then have an extra dose jabbed into them for good measure.

Luckily, the high nitrogen load in our environment doesn't need to be a long-term problem—provided we can find a way to end our emissions. There are armies of bacteria in the ground that get their energy from once-prized and now harmfully overabundant nitrogen compounds by breaking them down into their original components. When they do this, the gaseous form of nitrogen escapes from the ground and returns to its original home—the atmosphere—while another portion is washed down into groundwater by rainfall, spoiling our thirst for our most important form of sustenance. There is no doubt that the pendulum can swing back as soon as our interference in our ecosystem is reduced appropriately. And then, one day, salmon and bears will be in charge once again.

The full effect of this duo plays out only along streams and rivers, but there is another force of nature that makes its presence widely felt everywhere. It shapes mountains, forms valleys and wetlands, and is, most importantly, a gigantic redistribution mechanism: that force is water.

3

Creatures in Your Coffee

WATER NOT ONLY transports nutrients into the forest in the form of migrating fish, but, even more importantly, it also carries huge quantities out of there—thanks to its innate properties and the law of gravity. Water flows downhill. We all know this. But there's more to this seemingly mundane process than meets the eye. I'm talking about the survival of whole ecosystems.

Let's start by looking back into the past. All life on this planet needs nutrients such as minerals and compounds containing phosphorus and nitrogen. Nutrients dictate the vigor of plant growth, and all animals depend on plants for food. And this time, I'm not talking about salmon but about us. The ancestors of today's Central Europeans were part of a grand experiment that showed just how bound up we are in these cycles of life. First, they cut down forests to provide space and building materials for settlements. Then, the settlers farmed the ground they had cleared.

At first, the system worked well, thanks to the many tens of thousands of tons of carbon dioxide stored per square mile of ground in the form of humus. This soft brown material now began to decompose slowly. Without the cooling shade of trees, the ground warmed up, and bacteria and fungi became active even deep below the soil surface. In the orgy of consumption that ensued, not only was carbon dioxide exhaled into the atmosphere, but nutrients that had previously been bound up in the soil were also released. The overfertilization of the soil that resulted was welcomed at the time. Abundant yields ensured people were well fed even when other resources were scarce, and they continued for a few golden years until the fertility of the soil gradually diminished. As there were no synthetic fertilizers at the time and the paltry number of livestock produced only meager quantities of manure, eventually nutrient levels in the fields were depleted.

The soil still had enough nutrients, however, to grow grass. And so fields previously used to grow crops were now used to graze livestock. Of course, even in this scenario, nutrients were still being removed from the land, because the animals didn't remain in the fields: they were led away to be slaughtered and eaten. The land, therefore, continued to lose its vitality. Heathers and junipers—plants that sheep and goats don't eat—claimed more and more space. Eventually, all that was left were ruined fields that contributed next to nothing in the way of food. Today, we find such landscapes romantic: on a summer's day, there's nothing quite like a juniper heath or an expanse of heather dotted with sheep. For our ancestors, however, the sight of juniper berries or heather in bloom signaled destitution.

After the invention of synthetic fertilizers, large expanses of heathland were brought back into agricultural production, because farmers could now spread as many nutrients as they wanted. The few small areas that remained as a testament to old-time agricultural mismanagement are preserved and maintained to this day, but that is another story. What our ancestors did was participate in a grand experiment in speeding up time. They accelerated the natural emission of nutrients and inadvertently demonstrated what happens when there's no process to replenish them.

It's not that I long for the return of the days before fertilizers, because that would mean fully integrating ourselves into these cycles once again, and my father made it clear to me what that meant. In the postwar years, his family maintained a vegetable garden as an important source of extra food. Manure was scarce, so they spread the contents of the home septic system over the vegetable beds. This domestic fertilizer came back later processed into the salad greens and cucumbers that were served up at the table, enriched with an incidental gift from nature: intestinal worms. The worms cycled along with the nutrients from the toilet into the garden and back onto the table. But even such unappetizing results cannot prevent these kinds of nutrient cycles from slowly drying up.

And that brings us back to water. Water is a solvent, and it dissolves all the important substances plants like to suck up with their roots. This means that even though nutrients are taken up out of the ground by plants, they return to the ground when plants die and bacteria and fungi break them down. At least that's the simple version.

In the normal course of events, moisture seeps deep underground until it reaches groundwater. And on its way down, it takes with it all those vital substances that Trees & Co. would so love to keep for themselves. (That is, incidentally, also why our drinking water needs to be increasingly chlorinated—because the liquid manure that is spread on fields and pastures in unbelievable quantities also ends up, along with its generous portion of bacteria, many layers down in aquifers and, therefore, in our most important form of sustenance.) This natural vertical movement is of utmost importance for the ecosystem under our feet. Numerous creatures living deep underground depend on scraps from the table of life above the surface.

Before we turn to these beings, however, I would like to comment on the destructive power of water. Not every shower of rain seeps gently into the porous soil of the forest floor to replenish the groundwater. In heavy storms, pores in the soil fill up, and the soil's natural vertical channels overflow. When the ground is saturated by heavy rain, brownish runoff flows into the nearest stream, carrying with it a great deal of organic matter. You can see this for yourself every time you take a walk in the rain. As soon as the rivulets in pastures and fields become murky, they are carrying off soil—valuable soil that won't be replaced for a very long time. Sooner or later, the ground gets worn out as soil is washed away.

Or, at least, that's what would happen, but luckily nature has stepped in to stop this process of erosion. Nature's main line of defense is the forest. Trees slow the downpour by intercepting a great deal of the rain in the forest canopy. The intercepted moisture drips slowly down to the ground after the shower has passed. And that's why we Germans have the saying that it always rains twice in the forest. These leafy interceptors ensure that even heavy rainfall is spread out over a wide area and reaches the ground slowly so that the soil has time to absorb almost all of the moisture. Soft moss on tree trunks and ancient stumps helps by soaking up the surplus. These green cushions can store many times their weight in water, which they gradually release back into the surrounding forest. And because there's hardly any erosion when rain is slowed down like this, the soil layer in ancient forests is usually very porous and deep. It acts like an enormous sponge that absorbs and stores large quantities of water. Thus, intact forests create and protect their own reservoirs.

Without trees, the situation changes dramatically. Although grasslands can still somewhat reduce the impact of heavy rainfall, plowed fields have no protection against raindrops pelting down. The fine crumb structure of the soil is destroyed, and the pores fill with mud. Many crops, such as corn, potatoes, and turnips, cover the ground for only a few months, which means that fields are left completely unprotected from the weather for the rest of the year—a situation that nature did not anticipate in these latitudes. When a cloudburst hammers the ground, barely any water seeps downward; instead, floodwater streams over the surface.

The term "flood" is not an exaggeration here. A heavy storm cloud can rain down 100,000 cubic yards of water per square mile—in just a few minutes. If things don't run smoothly—that is to say if the rain is not slowed down by leafy growth so that it can seep into open pores in the soil—raging torrents quickly form and carve deep grooves in the muddy fields. The steeper the slope, the faster the torrents flow, and the more soil they wash away. A grade of just 2 percent, which looks flat as a board to us, is enough for soil to be lost—and the losses are dramatic.

Have you ever wondered why archaeological finds always have to be dug up out of the ground? Shouldn't they be lying on the surface, all grown over by grass or undergrowth? And why aren't mountains getting higher all the time? They are created when continental plates collide and thrust upward at the point of impact—a process that continues unaltered to this day.

Mountains, however, are not constantly increasing in height, for the same reason that Roman coins are usually found buried deep in the ground: erosion. Land is higher than the ocean (another blindingly obvious fact), and rain clouds formed over the ocean provide the land with a constant supply of water. The water flows downhill and arrives, sooner or later, back where it started. It picks up particles as it goes, imperceptibly scraping soil off the mountains. The steeper the terrain, the faster the water flows, and the more extreme the abrasion. Our landscapes, however, are not formed by steady, normal rainfall and peacefully babbling brooks but by rare extreme weather events. When rain pours down for weeks, turning small streams into raging rivers, then the mountains really take a battering. The resulting floods can shift even large rocks and carry away so much soil that the murky water turns a light shade of brown.

Once things have calmed down again, you can see the new contours of riverbanks where the water has undercut the river's sloping sides with particular force. As the river returns to its regular course, the receding floodwater deposits a thin layer of mud over the rest of the valley. The mud is a combination of water and dust, and the dust is abraded rock: bits of the mountain have been washed down into the valley. Valleys are fertilized by this sediment-rich floodwater—the Nile is a prime example. The highly developed civilization of ancient Egypt was possible only because fertile riverbanks allowed farmers to produce a lot of surplus food, and plenty of food on the table means plenty of time on your hands to invest in other activities.

BACK TO THE forest. What's happening here is Peter is being robbed to pay Paul, and in this case, both Peter and Paul are trees. Despite having a fondness for nutrient-rich soil many feet deep, many trees grow up high in the mountains. But the higher the elevation, the steeper the slopes, and therefore the more severely the soil erodes. And that is why trees on upper slopes don't grow as tall as trees lower down. The trees struggle mightily to fortify themselves against these forces of nature, and over time every crumb of soil they manage to hold on to makes a difference. Just four one-hundredths of an inch of erosion amounts to the loss of almost 4.5 tons of soil per acre. Agricultural fields in Central Europe lose an average of 1 ton of soil per acre per year; that's a loss of almost an inch over one hundred years.

In extreme cases, however, up to 20 inches of soil can disappear in that amount of time, and I can see the long-term consequences for trees in the forest I manage. There's a small hill here with an ancient beech forest on one of its slopes. Despite the steepness of the slope, there are at least 6 feet of rich soil on this side of the hill. I know the exact depth, because this is where, to protect a stand of ancient trees from logging, we established a cemetery in the forest—our Final Forest. And to establish the cemetery, we had to determine the soil's "interability," as it states in the red tape. To put this in words you can understand: Is it possible to bury urns here to a depth of 30 inches? We hired a geologist to find out, and, much to our astonishment, he came across this thick layer of soil. As he explained it, "The forest must have been here for a very long time"—probably since the arrival of the beeches about four thousand years ago.

On the other side of the hill, in contrast, bare rock is exposed in places. The once-thick soil has disappeared, leaving only a thin layer a couple of inches deep. Clearly, in the Middle Ages, pastoral farming happened here, and even though grassland performs better in terms of erosion than cropland, the results were disastrous. A few fractions of an inch of erosion over the ensuing centuries amounted to yards of soil lost and washed down into the neighboring Armuthsbach.

We now had a better idea of how the stream got its name. Without soil or humus, the fertility of the land was drastically reduced, which resulted in famine. Indeed, as recently as 1870, people here were dying of malnutrition, and covered wagon trains from Cologne had to bring food to the suffering villagers. The wagon trains were regularly ambushed by outlaws; it was like the Wild West out there. And all this happened because forests were felled, which then led to the almost imperceptible but inexorable erosion of the soil.

Can the process be reversed? Yes, it can. This is reassuring news, even if the timelines are as massive as those of the initial damage. Let's assume that the damaged ground is reforested one day and erosion basically stops: then the layers of soil will begin to build up again. As soon as the rate at which the soil is being eroded drops below the rate at which new soil is being created, the amount of brown gold will increase. The source of this new soil is rock, which is constantly being weathered slowly into tiny pieces.

In the conditions we have in Central Europe, on average 850 to 2,900 tons of rock are transformed into soil per square mile per year. That means an increase in soil depth of one one-hundredth to four one-hundredths of an inch, which would average at least 2 inches per century. It would take about ten thousand years for the rocky slope on the hill by the Armuthsbach in the forest I manage to return to the condition it was in before the trees were felled and the land was put to agricultural use—that is the length of time from the last ice age until today.

Does that seem disturbingly slow to you? Well, nature has time, as you can see if you consider the growth rate of trees. The oldest spruce in the world, in the Swedish county of

Dalarna, is almost ten thousand years old. Looked at that way, one generation in tree terms is all it will take until everything is back on track.

IN OUR SEARCH for ecosystems and their interconnections, we've already had a good look around on land. But, just a moment, that's not quite right. We've looked around above ground, but what's going on down below? The world is a three-dimensional place, after all, and yes, there are additional sizeable environments in the layers underneath our feet. I'm not talking about the 6 feet of arable soil I've just described, either. This time, I'd like to take you down far deeper. For bacteria, viruses, and fungi have been found at depths of up to 2 miles, and if you go down 1,600 feet, you can find many millions of these life-forms per cubic inch of matter. In these lightless depths, oxygen no longer plays a role in breathing, and in many cases food consists of the materials we like to use for industry and transportation: oil, gas, and coal.

Life in these hidden ecosystems has barely been explored, and we know only a tiny fraction of the species that live down here. According to the first rough estimates, the rock layers could be home to 10 percent of the earth's total living biomass, and as they are way down deep and mostly inaccessible, we can assume that, apart from a few coal mines and deep strip mines, these layers have been spared huge disruptions by human activity.

These depths hide another distribution system, one where humans have begun to interfere: groundwater. Groundwater is a very special habitat. Not a single ray of light ever reaches down here, and neither does cold. Depending on the depth, it's either pleasantly warm or extremely hot, and there's not much to eat. In times of climate change, these ecosystems offer a distinct advantage—down here nothing changes.

Despite the lack of food, there's some lively activity going on beneath our feet. Okay, maybe not that lively, because—at least in the layers close to the surface, where the temperature sometimes drops below 50 degrees Fahrenheit—it's not particularly warm. Low temperatures and little food mean animals slow down. When you get to 100 to 130 feet below the surface, the temperature rises to 52 to 54 degrees, and it increases by 5.5 degrees every 300 or so feet as you continue to descend. However, if you believe that life is lived at a faster pace down here where it's warmer, then you're mistaken.

Topping the charts of the slowest creatures ever is one of the most happy-to-reproduce life-forms in the world: bacteria. Whereas many members of this group reproduce at a breathtaking rate (in our gut, for example, some species of bacteria divide, that is to say double their numbers, every twenty minutes), it seems that inhabitants of the layers buried a half mile or more below the earth's surface are immune to the pressures of time. As reported in *Der Spiegel* online after the 2013 fall meeting of the American Geophysical Union, some species take five hundred years to divide.[1] In these conditions, food doesn't spoil and no bacterial diseases break out, because the hosts (us) would be dead long before the diminutive creatures had even begun their work. The slow pace of life is because of the inhospitable conditions: at these depths, high pressure and high heat are the norm. The record holders among these tiny creatures so far survive 250 degrees Fahrenheit and keep merrily dividing—at their own pace, of course.

In this realm of the deep, it seems at first glance that little changes over the course of centuries. But that isn't quite true, because everything down here is in flux. After heavy rains, water seeps down from the surface—at least it does in Central European latitudes, where more freshwater falls from the sky every year than evaporates into the air again. If less rain fell, Central Europe would become a desert, and in some regions, it wouldn't take much to tip the scales—as a quick look at the numbers makes abundantly clear. In Germany, on average, 12 gallons of water evaporate each year—per square foot.[2] In some areas of the state of Brandenburg today, it hardly ever rains anymore, which means that the groundwater here is basically not being replenished. As climate change continues, the rate of evaporation will continue to increase, so it probably won't be long before the underground realm will finally be cut off from resupply from above, and the system needs the resupply, because a small amount of moisture is constantly being lost here and there.

Freshwater springs are groundwater's gaping "wounds." What we see as merrily bubbling wonders of nature are a complete catastrophe for some of the dwellers of the deep. Water forced up to the surface through layers of rock rudely washes crustaceans and worms into the light of day, where, thanks to the abrupt change in their environment, they quickly expire. Winter is an especially good time to identify these upwellings of groundwater, because they don't freeze where they bubble up out of the ground. (Groundwater maintains a constant temperature of about 50 degrees Fahrenheit, which drops only when it encounters fresh air and all around is frozen solid.) So, if you find gently moving, ice-free water when temperatures outside are well below freezing, you can be certain that the water is coming from deep below the surface.

Let's get back to species diversity. According to recent research, groundwater harbors a surprising wealth of crustaceans and other minuscule creatures. They paddle blindly through dark currents, and, every once in a while, they probably end up in the water you use to make your morning coffee. Most treatment plants pump water into their reservoirs from deep below the surface, tapping into what until their intrusion was basically a hermetically sealed habitat.

Tiny creatures in your coffee despite elaborate filters in water treatment plants? Yes, despite all efforts to keep them out, pesky little creatures such as water lice (which can grow to almost an inch long) often make it through to live happily in the water pipes on the other side of all those purification systems. After all, the water pipe in your basement is basically an extension of the groundwater system—it's dark here, clean, not too hot and not too cold.

You notice the connection the moment you turn on the cold water faucet: the water that comes out is the temperature of groundwater. When you start the flow, some of those little scoundrels lose their grip and are swept along in the stream of water—and end up in your stomach by way of your coffee. But water lice are not the only creatures in the water system;

many others are smaller still. Bacteria, for example, form a thick layer that coats the inside surfaces of the metal pipes. And there are traces of them too in every sip we take.

You can look as closely as you like, but you won't be able to see most of these uninvited guests (with the exception of giants, such as water lice), at least not without the aid of a microscope. In the absence of light, neither eyes nor color make any sense, and that's why the inhabitants of groundwater are typically blind and transparent. The lack of light, however, brings up another problem. Without sun, there's no photosynthesis, and so there's no food being produced by plants. The hordes of basement dwellers on this earth are dependent on handouts from the biomass of plants and animals above that decompose into humus and sink slowly into the depths along with the rainwater that seeps into the ground.

On the way down, the nutrients are metabolized multiple times, because there's an entire food chain down here, just as there is above ground. Most inhabitants are bacteria, which colonize everywhere and form layers, just as they do in basement water pipes. These swards of bacteria are browsed by minute predators, such as flagellates and ciliates. It's a good thing these voracious Lilliputians exist, because if they didn't, the pores of plutonic rock deep below the surface would eventually become clogged. But even these tiny beings meet their match: heliozoans, also known as sun animalcules. They are a little bit bigger and particularly enjoy eating their fellow creatures.[3] And so there is a complete ecosystem underground of which we are only dimly aware, except where we pump its (and our) elixir of life—water—up to the surface to satisfy our needs.

Speaking of which, we got sidetracked while we were having our morning coffee and contemplating the stowaways in our mug. If you are disgusted by the thought of bacteria in your beverages, perhaps this extra tidbit of information will make you feel better: you yourself are a mother ship for these tiny creatures. Apart from the 30 billion cells that make up your body, you are also host to about the same number of bacteria, most of them in your gut.[4] Thousands of different kinds of bacteria are floating around inside you, and in most cases, they are vital for your survival—for example, by helping you fight off illness or break down foods that are hard to digest. Does it really matter if a few more harmless fellows find their way inside you via your drinking water? They won't survive for long in your digestive tract anyway.

FORESTS ARE IMPORTANT for groundwater, so important that some water companies in Germany even pay forest owners bonuses for following responsible management practices—which sounds counterintuitive. First of all, trees are enormous consumers of water. For example, on a hot summer day, a thirsty mature beech can suck up to 130 gallons of water from the ground. It uses the water for a number of different purposes, but most of it ends up evaporating out of its stomata (the tiny openings on the underside of its leaves). Grass would use far less water.

But trees, especially deciduous trees native to Central Europe, have an advantage over grass because they also collect water. Their upward spreading branches gather cool rain and funnel the water to the trunk and down to the roots. I once stood under an ancient beech in a heavy storm (don't try this at home!), and I observed this water gathering for myself. So much water shot down the trunk that it foamed up around the base of the tree like freshly poured beer.

Once the water reaches the ground, it soaks into the loose soil, which is as absorbent as a sponge. Even heavy downpours soak in and slowly trickle down through the soil layers. It's true that trees help themselves to some of this water later when there's no more rain—for trees, the ground around their roots is like a reservoir that they can tap into anytime they're thirsty—but the rest seeps down to layers where plant roots can no longer reach it, because they don't grow that deep. And in the depths, the water slowly becomes part of the flow of groundwater.

Where I live, resupply of groundwater happens only in winter, when the plant world is hibernating. Beeches and oaks take a break for a while, and the water can slip past the tree roots unchecked into the depths. In summer, in contrast, there's never enough rain to satisfy the trees' thirst. They greedily suck all the moisture out of the ground and pump it into their trunks.

The way trees take up water gives me pause for thought in these times of climate change. Warmer temperatures will change many of the parameters. Water will evaporate more quickly, which means that the ground will dry out sooner, even without plant activity. In addition, just like us, trees drink more in hot weather. And longer growing seasons will shorten the times when trees take a break and forests hibernate and the ground can recharge its supplies of water. But despite these issues, forests should still be able to create enough new groundwater beneath them in the future—as long as we don't damage them too much with our logging operations.

Open grassland and land cultivated for agriculture, however, are less capable of absorbing rainfall. Wild or domesticated grazing animals compact the surface layer. In modern times, large agricultural machinery is the main cause of compaction, and compaction by machine reaches far deeper than compaction by hoof or trotter. The spongelike soil is compressed and, unlike the sponge beside your kitchen sink, never regains its original absorbent structure. Heavy rain can no longer be absorbed; the water runs off downhill in rivulets that flow with increasing speed and end up in the nearest stream (which leads to the nearest river, which carries the freshwater down to the ocean). And so the water is lost to the local groundwater supply, and the whole process speeds up with erosion.

The air heats up much more quickly over pastures and fields than it does over forests, which means that the ground dries out more quickly and life-giving moisture escapes into the air, where it is carried away, which intensifies the effects of desiccation.

The greatest danger for groundwater, however, is not climate change but the extraction of raw materials, especially through fracking. In fracking, water is pumped deep into the ground under high pressure to fracture rocks. Grains of sand and chemicals mixed in with the water hold the fractures open, which allows the gas and oil contained in the rocks to flow to the surface. The underground ecosystem is not equipped to deal with such brutish intrusions. In this realm, after all, hardly anything ever changes, and the changes that do happen, happen extremely slowly. One can only hope that not too many areas are opened up for this method of mining.

Other than avoiding fracking, the forest is the best means of protecting groundwater. Its trees are the secret protectors of tiny crustaceans in layers 1,000 feet and more beneath their roots. There are other animals, however—deer, for example—that have a more strained relationship with beeches and oaks. You could say that the relationship between trees and deer literally leaves a bad taste in the deer's mouths, and, as it turns out, deer don't taste good to trees, either.

4

Why Deer Taste Bad to Trees

DEER HAVE A love-hate relationship with trees. Deer don't actually like forests, but we think of them as forest animals, because that's where we find them most often. Like all large animals that eat plants, deer have a problem: they can only eat vegetation they can reach. And usually, the vegetation available to them has armed itself against herbivorous attack. The usual arsenal of vegetative defensive weapons includes thorns and barbs, toxins, or thick, hard bark, but trees in Central European forests have developed none of these defenses.

Does that mean that the trees' offspring have to endure every bite from browsers without being able to fight back? If you take a good look around the forest, you will see how beeches defend themselves. The forest floor around deciduous trees is conspicuously empty of vegetation. Here and there, you might find a lonely fern or a few grasses in a tiny clearing where an ancient giant has fallen, allowing a few rays of sunlight to reach the ground. Light levels in general, however, are too low for plants to produce copious quantities of sugar, which means that wild plants growing in the forest contain few nutrients in comparison with their relatives growing out in the open. In other words, understory plants are tough and bitter.

Most of the forest is darker still, because only 3 percent of the sun's light penetrates the canopy. This makes it pitch dark for the plants under the trees. You might not think so when you walk through the trees, but it has to do with the green shade you find in the forest. Trees use chlorophyll in their leaves to convert light, water, and carbon dioxide into sugar. Chlorophyll, however, has a "green gap," and it cannot make use of this wavelength. Therefore, green light is reflected, which makes the forest seem brighter to human visitors than it does to plants, because plants—unlike us—cannot "see" this color. As 97 percent of all the other wavelengths of light have already been absorbed and processed in the canopy, from the point of view of the green plants growing on the forest floor, things literally look gloomy.

Young beeches belong to this group, of course. The few rays of sunlight that fall on their small thin leaves allow them to produce such meager amounts of sugar that their twigs and buds contain barely any nutrients. To ensure that the next generation of trees doesn't starve because its opportunity to photosynthesize is so limited, mother trees supply the saplings with nutrient solutions through connected root systems—you could say, they suckle their offspring. Nonwoody plants and grasses, however, don't benefit from this type of assistance and therefore can only grow in the tiny clearings opened up when an ancient giant falls.

So this is what the supposedly idyllic forest looks like to roe deer: a few patches of dry, stringy grasses and nonwoody plants scattered among tough young beeches. Even if the beech leaves were halfway palatable, they would make a very boring diet, and deer don't enjoy that any more than we do. Imagine having to eat your favorite food day in and day out for months on end—after a few days, you'd lose all taste for it. Deer prefer to avoid monotonous meals that don't offer much variety in either taste or nutrients, especially when they have to produce milk for their fawns. They find forest edge habitats much more attractive. Here, maybe along a riverbank, grasses and nonwoody plants grow in full sun in fertile soil until they positively brim with energy. Unfortunately, in forest-rich Central Europe, few edge habitats occur naturally, which is why historically forests here have had only very small populations of roe deer.

It's no surprise that roe deer prefer disturbed areas. When a summer tornado uproots a small stand of ancient beeches, an island of light opens up in the forest. The plants I have just described that struggle to survive in forest shade soon settle the clearing. And they have a lot to offer deer. Bright sunlight means full-on photosynthesis for plants, which means tasty carbohydrates in their leaves and buds. Even the beech saplings, which now find themselves unexpectedly illuminated, become sweet and tasty. And so, here in the clearing is the paradise the smallest species of deer in Central Europe has been searching for.

Roe deer love food with a high energy content; scientists call them concentrated feeders. If we were to eat the same way roe deer eat, our meals would consist of nothing but fast food and chocolate enriched with vitamins. Roe deer don't have to worry about getting fat, however, because, as I just mentioned, such calorie-stuffed islands rarely occur in nature.

If you're a small plant eater, it's a bad idea to run in the face of danger. Wolves could easily follow you and attack, so it's better for you to hide. Roe deer don't run very far before they double back and try to return to their original location. When they do, they cross their own tracks, which confuses their pursuers—which trail should they follow? Once they're safely home, roe deer hide in stands of small trees. And because herds are easier to spot than single animals, roe deer live alone. Another reason for their solitary existence is the lack of food in ancient undisturbed forests. A herd of deer would have to cover a lot of territory to find even small quantities of food. Traveling long distances, however, increases the risk of coming across a pack of wolves. And so the single life is better.

The need to be solitary also means that a mother leaves her offspring behind when she goes out in search of food. This is perfectly normal behavior in the first three to four weeks after a fawn is born—the time when the little one can't yet keep up. So that her fawns (usually she has twins) don't slow her down, the doe leaves them lying in tall grass or under a bush. When an enemy approaches, they flatten themselves to the ground so that they can't be seen. Unfortunately, some people interpret this behavior to mean that a fawn has been abandoned, and they take the supposedly helpless young deer home with them, where it often dies a painful death from starvation because it refuses milk from a bottle.

Life without a large extended family is typical for many forest dwellers, including, for example, the lynx. Lynx strike out alone through their enormous territory, which can sometimes exceed 40 square miles, and they seek brief contact with a lynx of the opposite sex only at mating time. By contrast, red deer, which originally lived on grassy plains, behave completely differently. They are social animals that live in large herds, and they only go off on their own when the does are ready to give birth—which every red deer doe likes to do in peace and isolation. When predators appear, red deer flee together over great distances in search of a place with good visibility in all directions. They've retained this behavior, despite the fact that human activity—in Central Europe, at least—has forced them to retreat into our modern-day forests. We don't want to share our open spaces with them, because these are the places where we like to settle and farm.

Back to the roe deer. Life is better for them than ever in Germany now that no dark ancient forest remains. What we think of as forests today are drastically different from forests of the past. If you get a bird's eye view of the landscape—perhaps by taking a look at a satellite image on the internet—it looks like an enormous patchwork with pieces missing, and the forest patches are small from an ecological perspective, because anything less than 80 square miles is not large enough to support even one wolf pack.

The many tiny scraps of forest are a huge advantage for roe deer, because they can find their preferred edge habitats all over the place. Whereas once the collapse of a number of trees gave them a lucky break, these days, enough light reaches all over the forest floor so that nonwoody plants and grasses proliferate inside as well as at the edges. Forestry, after all, is simply the practice of growing and cutting down trees. Clear-cuts are the most brutal form of harvesting timber, yet they are a windfall for browsers. Once the pesky shade from tree-tops has been removed, nonwoody plants and grasses can take over. Not only do the plants now have space and light, they also get a massive boost of fertilizer. The bright sunlight heats the ground so much that fungi and bacteria even deep below the surface warm up enough to spring into action and break down all the humus in less than a few years. So many nutrients are released that even the exploding plant populations cannot take them all up. The plants grow quickly and are full of sugar and other carbohydrates, making them tasty treats for roe deer. In areas such as these, roe deer don't need to move around much: they can find enough food in a few square yards to keep them comfortably full for the whole day.

Under such circumstances, populations of plant eaters explode, because, like all species, they immediately convert food into offspring. Instead of one fawn, there are two or perhaps even three, and the ratio of the sexes moves in favor of females. That heats up population growth even more, which is ideal from the animals' perspective, for it means that roe deer can completely take over the habitat and exploit it down to the last blade of grass.

There have been noticeable increases in the growth of wild animal populations in Germany, especially after severe storms like those in 1990 (Vivian and Wiebke) and 2007 (Kyrill), which mowed down entire forests. Spruce trees bore the brunt of the damage, but pines and Douglas firs growing in plantations also suffered. The conifers began to tumble when winds blew more than 60 miles an hour. The trees fell because their root systems were damaged when they were still in tree nurseries. Staff cut their roots back to make them easier to transplant: if the roots are shorter, then you don't have to dig such a big hole for the trees. The flip side of this coin is that transplants with trimmed roots never develop an intact root system, which means it's almost impossible for them to hold fast to the ground when storms hit. What makes things worse is that conifers hold on to their needles and therefore have lots of surface area to offer winter storms. This is in marked contrast to beeches and oaks, which pare down their profile by dropping their leaves in the fall and therefore survive most winter storms unscathed. And so it is that conifer plantations indirectly favor roe deer.

In the past, in addition to storm damage, we also had the intentional clear-cutting of trees as part of commercial forest practice. With a clear-cut, you harvest a complete stand of trees of the same age in a single pass, which is much cheaper than felling individual trees in thinning cuts. Recently, however, clear-cuts larger than 2.5 acres have fallen out of fashion. Too bad for the roe deer? Not at all, because thinning cuts are almost as beneficial for vegetation on the forest floor. Individual trees are removed regularly to give particularly good specimens room to grow. Constantly removing trees in a regular rotation is just making a more widely distributed, less intense clear-cut. In contrast to an untouched ancient forest, trees make up less than 50 percent of the biomass of a cultivated forest. This means that more light reaches the ground, nonwoody plants, grasses, and bushes take hold over large areas, and the lower stories of the forest become warmer (by about 5.5 degrees Fahrenheit). The resulting deer buffet is not as lavish as the feast laid out in the clear-cut; however, it makes up for this by being available almost everywhere in the forest.

As about 98 percent of the forested areas in Germany are cultivated, that's like having a gigantic nationwide feeding program for deer. Add to that the hunters who put a great deal of effort into caring for their potential prey by hauling tons of feed into the forests, and you can see why prey populations are increasing by leaps and bounds. Today, as I mentioned earlier, there are fifty times more roe deer wandering through our forests than there were before these intrusive measures started.

You can easily see for yourself where and how forest landscapes in Central Europe have changed. With the exception of a few tiny areas, grasses, nonwoody plants, and bushes hardly ever grow in natural forests in these latitudes. Large-scale coverage by these plants can always be traced back to human interference in this ecosystem, which for roe deer, at least, is cause for celebration.

Some of the plants see things differently, because, when it comes to food, this little species of deer has preferences, just as we do. At the top of the list are saplings of beech, oak, cherry, and other deciduous trees, along with the offspring of the now-rare white pine. After that, the 3-foot-tall stalks of fireweed topped with spikes of radiant magenta flowers and the somewhat inconspicuous wild raspberries are perennial favorites. These are the first delicacies to be eaten, and if roe deer populations are high, these plants disappear completely, allowing others better able to defend themselves, such as blackberries, thistles, and stinging nettles, to take over.

It's easy to deduce that trees native to the ancient forests of Central Europe never had to deal with browsers: they have developed next to no defenses against the hungry mammals. No thorns, no toxins in their leaves, no impenetrable tangle of branches. No, beeches and oaks offer their almost defenseless seedlings to any animal that might want to take a bite out of them. Their only protection is the eternal twilight of the forest floor and the aforementioned lack of plants, which make the forest an unattractive place to live.

These weak attempts at defense, however, work only when there are very few animals around, as is the case with roe deer, for instance. These defenses would not work against large herds of hungry aurochs or tarpans (ancient horses), which would simply have torn the bark off the trees. The dying trunks and crowns would have created space and light for grassy plains to get established, and the plant eaters would then have been able to nourish themselves from plants that grew out in the open—and the forest would have disappeared.

But none of that happened in Central Europe. As far as I'm concerned, that is a clear indication that there never was a serious, long-term threat from these animals, or evolution would have come up with countermeasures.

Things are quite different for plants adapted to life on grassy plains. Wild horses, wild cattle, and red deer are all at home on these wide-open spaces, and they love to nibble on the fresh growth of trees and bushes to vary their diet. Woody species that grow in such surroundings defend themselves vigorously against their attackers. Blackthorn is a classic example. The dagger-like thorns—even on bushes that have been dead for years—are sharp enough to easily pierce not only skin (of any kind) but also rubber boots and car tires. The wild apple deploys similar defensive weaponry. Like the blackthorn, it also belongs to the rose family. Roses = thorns = grassy plains.

Plants that prefer not to equip themselves with thorns depend on toxins. Among these are foxglove, broom, and ragwort. This last one is especially dangerous because its harmful effects accumulate over time. First it causes mild liver damage, but at some point the animal eats one plant too many and it dies.

Not every species, however, is affected by the ragwort's poison. There are butterflies and moths that not only consume these pretty yellow-flowering plants but also use them for their own protection. The cinnabar moth is one example. Cinnabar moth caterpillars happily chew through one tiny leaf after another all day long, taking in not only the calories the plant provides but also its poisons. The caterpillars suffer no ill effects whatsoever, but the predators that eat them do, and the caterpillars sport black and yellow rings to warn attackers that they will make a deadly meal. This color combination seems to be a universal warning in the animal kingdom: just think of the yellow and black on wasps and salamanders, for example.

Throughout the landscape, plants struggle not to be eaten. Although they are very quiet about it, recent research has shown that deciduous trees are not as passive as we (and I) long supposed. To find out more about this, scientists at Leipzig University and the German Centre for Integrative Biodiversity Research (iDiv) simulated attacks on tiny beeches and maples. Whenever a roe deer takes a hearty bite out of the top growth of a young tree, it leaves a little saliva behind in the wound, and it soon became clear that wounded trees can detect the presence of this saliva. To simulate browsing by roe deer, researchers cut off buds or leaves and dripped roe deer saliva from a pipette onto the damaged areas. What they noticed was that the little trees produced salicylic acid in response, which ensured that the trees increased their production of bad-tasting defensive compounds, which discouraged the roe deer from eating them. When, however, the scientists simply broke off new growth without applying any saliva, all the beeches and maples produced were hormones to heal the damage as quickly as possible.[1] This experiment also proved that these trees (and possibly many other species) can "taste" the saliva left on leaves and shoots and are aware that they are being browsed by herbivores.

After the deer population reaches a certain density, however, knowing what's eating you doesn't help. The animals polish off so many of the plants in their territory that they will even nibble all the bad-tasting leaves off beech saplings. Forest owners at their wits end try to help the little deciduous trees by smearing bitter-tasting concoctions on their leaf buds. Even I tried this in my early years as a forester, but the roe deer soon put paid to this preventative measure: they were so hungry that they just ate the white paste along with the buds.

Forest floors scoured of vegetation—and the ageing of the forest that ensues—is an acute problem in many areas in Central Europe, and it shows that wild animals have reached population levels that the trees have never had to deal with before. How might we change this dynamic in the future? One way would be to leave more trees in the forest. In other words, foresters could step back. Allowing more trees to grow would make it darker once again, which, in turn, would allow beeches and oaks to employ their tried-and-true strategy of light starvation. And if hunters also gave up their winter feeding programs, the situation would improve considerably. If the wolf were to arrive as well (and it's on its way), perhaps we'd eventually have a situation similar to Yellowstone here where I live.

None of this would make the clockwork of nature completely regain its former rhythm, because no one can and no one will remove the patchwork of pastures, agricultural fields, and smaller forested areas that covers the landscape of Central Europe—not even me. After all, I, too, am hungry every morning, and I enjoy biting into my breakfast roll as much as anyone else, and for that I need someone to farm a wheat field.

But it is not only roe deer that profit from our transformation of the landscape to suit our own goals. There are other brown animals that have a great deal of influence on the environment in which we live: they are teeny tiny, really good at defending themselves, and they have a weakness for forget-me-nots.

5

Ants—Secret Sovereigns

ALL SUMMER LONG, our yard is full of multitudes of forget-me-nots. The patches of blue pop up here, there, and everywhere, and are always wandering uninvited into our vegetable beds, where they make themselves at home and refuse to leave. And because they are so pretty, we usually leave them alone and accept their intrusion. Forget-me-nots, however, can only conquer new territory so successfully because they have an army of tiny allies: ants.

It's not that ants are particularly fond of flowers—at least, they are not attracted by their aesthetic qualities. Ants are motived by their desire to eat them, and their interest is triggered when forget-me-nots form their seeds. The seeds are designed to make an ant's mouth water, for attached to the outside is a fleshy structure called an elaiosome, which looks like a tiny cake crumb. This fat- and sugar-rich morsel is like chips and chocolate to an ant. The tiny creatures quickly carry the seeds back to their nest, where the colony is waiting eagerly in the tunnels for the calorie boost. The tasty treat is nibbled off and the seed itself is discarded. Along come the trash collectors in the form of worker ants, which dispose of the seeds in the neighborhood—carting them up to 200 feet away from home base. Wild strawberries and wood violets also benefit from this distribution service: ants in nature's employ as gardeners, as it were.

There's a huge army of them puttering around in forests and fields, and in some respects they are as busy as we are. About ten thousand species of ants have been discovered to date, and the German national weekly newspaper *Die Zeit* once took the time to estimate the total weight of all the tiny creatures in this insect family. According to their calculations, their combined weight is equivalent to the weight of all the people on Earth.[1]

Although most wood ants are small, their colonies and the structures they build are often very large. The largest anthill I've ever found in the forest I manage is almost 16 feet wide. My first experiences with red wood ants (which are the most common forest ants) were on childhood walks with my family. As soon as one of us spotted a sizeable hill of these social insects at the edge of the path, the ritual was always the same. My mother positioned herself next to the structure and tapped the outside gently with the palm of her hand. Then we were allowed to smell her hand—and we immediately detected a pungent, sour odor. Turning to present their rear ends, the ants had sprayed acid to fend off the intruder. During the demonstration, we had to hop quickly from one foot to the other to make sure none of the feisty little creatures ventured over our shoes and up our pant legs to bite us—their bites are extremely painful.

It's no surprise that wood ants are so good at defending themselves. After all, they are related to honeybees, and they share a similar social organization—except that ant colonies can have multiple queens. Also, related ant colonies tolerate each other, which can't be said of honeybees. The latter raid each other's hives, especially in the fall, and the defeated colony is mercilessly dispatched and its hive emptied of its honey. Ants are more peaceable—at least when it comes to dealing with their own kind—and they like other insects, but only in a culinary way. Adult bark beetles and larvae, for example, are eagerly carried away and fed to the ant larvae back at the colony. They are so insatiable that during the summer, millions of beetles in a radius of up to 150 feet or so from the hill end up as ant meals.

In spruce plantations, the dreaded spruce engraver beetles feed. In large monocultures of pines, it is the larvae of pine-tree lappets and pine beauties that strip whole forests bare. Not, however, in the vicinity of the hills of red wood ants. Here in the colony's home range, islands of green remain in an ocean of dead trunks. This quickly led to red wood ants being dubbed the public health patrol of the forest. Henceforth, the ants working so hard on behalf of foresters and forest owners came under strict protection, for red wood ants eat not only the aforementioned pest species but also carrion, which makes their title even more appropriate. Even though it was not their intent, the new rules also benefit rare species of birds that dine on ants. Woodpeckers such as the crow-sized black woodpecker, along with black grouse and wood grouse, or capercaillie, like to help themselves to larvae and pupae from anthills. Red wood ants, therefore, can clearly be categorized as beneficial.

But if we take another, closer look at the species, mild doubts arise. For example, there's the question of whether these ants are indeed worth protecting. I want to be quite clear here. Every species, regardless of whether it is common or rare, is worth protecting, in the sense of showing it respect. However, worth protecting, in the sense of mandating active support for the animal, is something altogether different, and something that in the ants' case is misguided, at least where I live. Red wood ants arrived as a result of changes we made, and they are able to expand their range only because of the unrestrained expansion of conifer plantations. These builders of hills were not represented in the original deciduous forests of Central Europe. Have you ever seen a red wood ant hill made out of leaves? No, these ants only use needles. Moreover, they need a lot of sun to be able to start work in the spring. The ants go to the surface of their hill to heat up in the sun, and then they crawl back inside to radiate the warmth they have gathered. Sun that reaches the ground would have been a scarce commodity in our original beech forests—yet another strike against the tiny construction engineers.

But even in their natural habitat, it is questionable whether the effect red wood ants have is totally beneficial for the trees. The trees are no doubt happy that the insects remove the bark beetles that attack them, but the ants' diet is not restricted to meat. It includes sugary foods, as well. And in the forest, sugary treats come almost exclusively from aphids. Aphids attach themselves to the trees' needles and bark, stick their mouthparts down to where the trees' sap flows, and tap into the trees' lifeblood. Thanks to photosynthesis, this "tree blood" has a high sugar content, but that's not what the aphids are after. What they want is protein, which is found in this fluid in only very small quantities. Therefore, the aphids need to allow enormous amounts of the trees' fluids to flow through their bodies so that they can filter out enough of the scarce substances they desire.

Whoever drinks a lot must also excrete a lot, and aphids excrete almost constantly. If you park under aphid-infested trees in summer, your windshield will tell you all you need to know—in just a few hours, it will be covered with sticky droplets. And because the little creatures are constantly eating and excreting, over time their rear ends can get gummed up with sugar. Some species resort to covering their excretions with wax so that they can expel them more easily; others enlist the help of ants. Ants lap up the sugary feces, because, like their relatives the honeybees, sugar is the most important component in their diet. Per season, a single ant colony digests about 50 gallons of these sugary droplets. This so-called honeydew makes up two-thirds of their calorie intake. To give you an idea of how much honeydew this is, an average of 10 million insects weighing a total of 60 pounds end up in the ants' stomachs to make up the remaining third. The small portion left is provided by tree sap and fungal threads.[2]

Wood ants and aphids, therefore, come as a package, and this is where the term "public health patrol" starts to slip, because aphids damage trees in many different ways. First, they sap beeches, oaks, and spruce of energy these trees desperately need for themselves. Then, the aphids inflict heavy damage on the trees' tissue when they insert their mouthparts and begin siphoning off fluids. The red-eyed nymphs of the spruce aphid, for example, which are less than one-twelfth of an inch long, drain the needles of many different species of spruce. The needles turn yellow, then brown, and finally fall off. Afterward, the trees look as though they've been plucked, because only the new season's growth remains on their branches, and their growth is stunted, because the trees' options for photosynthesis are now severely limited.

To these constraints are now added pathogens that can be deadly for trees. A kind of aphid known as beech scale feeds on the bark of beeches. These small creatures are covered in soft waxy hairs. They're not dangerous, as long as there aren't too many of them—beeches have no problem healing individual small puncture wounds. The situation is quite different, however, when their populations explode. Aphids have no need of males to multiply, and none has ever been found. The females lay unfertilized eggs that hatch into larvae. These larvae are carried on the wind to nearby beeches, where they immediately begin feeding. When all the nooks and crannies in the bark are occupied with colonies of white scale, the trees look as though they're covered in a light layer of mold, and the defense systems of many of them become overwhelmed. The tubelike mouthparts of the scale create weeping wounds that won't heal. The sap that bleeds out is colonized by fungi, which eventually work their way into the trunks and kill the beeches. A good number of trees survive the infection, but they carry the scars on their bark for the rest of their lives.

Loss of life-giving energy and the spread of disease through puncture wounds—aphids really are no blessing for the trees. And now the so-called public health patrol turns up. The red wood ants could just gobble down the little green pests and bulk up on the protein they provide. However, it is clearly far more advantageous for them to keep the aphids around and play honeydew farmer. They have to harvest 50 gallons of honeydew somehow, and what better way than by keeping aphids in the trees around the anthill? The ants benefit twice as they defend their herds of aphids against predators: not only do they get to protect their source of honeydew, but they also get access to prey such as aphid-loving ladybug larvae, which come to eat aphids but end up getting eaten by the aphids' protectors.

Despite the protection the ants offer, however, the aphids are not always content to stick around. When they want to leave, the next generation grows wings so that they can fly to greener pastures. This doesn't escape the notice of their guardians, and the ants end the aphids' dreams of flight by summarily biting off their transparent appendages. And as if that were not enough, the ants also use chemical means to prevent their domesticated herds from escaping. The ants exude compounds that slow the growth of the aphids' wings, and, for good measure, they also slow down the aphids: a research team from Imperial College London discovered that aphids move more slowly when they cross terrain that has previously been walked over by ants. The cause for the slowdown is a chemical message left by the ants that affects the behavior of the aphids and forces them to reduce speed.[3] The beautiful symbiotic relationship between ants and aphids turns out to be not entirely voluntary, after all.

Now you might argue that the aphids still benefit from the ants' attentions: for example, they don't have to worry about being attacked by ladybug or hoverfly larvae. And the "milking" process doesn't harm them. After all, the sugary droplets are simply waste products—and they are even being carted away to keep the aphids nice and clean. The sticking point is that the aphids would rather seek out more productive trees when they notice that conditions are no longer optimal in the place where they originally landed. But their protectors, who have now turned out to be their jailers, prevent them from moving away. Jailer ants that keep their "livestock" in trees in unnaturally high concentrations are supposed to be the public health patrol?

Are red wood ants really helping foresters when they weaken the trees around their anthills by setting up their honeydew farms? It's not an easy question to answer. At the beginning of this chapter, I mentioned the green islands left in coniferous forests after bark beetles have invaded. No matter how many aphids live in the trees that have been saved, the live trees are still better off than their dead companions. And this brings us to the key to understanding the complicated coexistence of different groups of insects. Trees are attacked not only by aphids and bark beetles but also by a multitude of other species, all of them with one thing on their mind: getting their share from the gigantic warehouse of carbohydrates that is a tree. Woodboring beetles lay their eggs on bark and their larvae then tunnel into it. Weevils chew leaves until the edges look as though they've been riddled with shot. This kind of damage is probably much more detrimental to the trees than donating some of their vital fluids to aphids. Red wood ants certainly ensure that there are more aphids around—which means more loss of blood for the trees—but that also leads to more ants in the neighborhood. Lots of food in the form of fluids = lots of ant larvae that can be fed. And the more the ants climb up into the trees to hunt the insects that threaten their aphid herds, the fewer attacks from these predatory insects the trees have to endure.

The more interesting question is what the overall ant-aphid-forest balance looks like. Science hasn't yet come up with a definitive answer; however, most studies conclude that the positive effects generally outweigh the negative ones. John Whittaker at Lancaster University, for example, discovered that, on balance, birches did significantly better when there were ants around. Although ants increased the number of aphids, this was true only for some species. Species of aphids not farmed by ants declined drastically. Furthermore, leaf-eating insects in general declined so significantly that leaf loss was six times less than in birches not settled by ants.[4] According to Whittaker, plane trees also seem to come out mostly ahead. Aphid-farming ants reduce attacks by other plant-eating insects to such an extent that plane trees with ants increase their girth two to three times faster than plane trees that have to manage without the ants' protection.[5]

Does that mean that wood ants are beneficial? I believe that the ecosystem is too complex for us to be able to answer this question definitively. If we take this one step further, you will see that trying to understand all the connections here is a Sisyphean task. We could start by asking about sugar. At the end of the day, despite the aphids' bloodletting, a tree can always produce more, because it still has leaves that are no longer being eaten by caterpillars. However, the sugar would normally remain inside the tree, and from there it would end up in the soil ecosystem via the tree's roots and the fungal networks in the ground, but thanks to the multitude of well-tended aphids, sugar from inside the tree now drips down onto the ground and the vegetation below. The hill builders can't consume all of it quickly enough, so many leftover droplets land on leaves and the soil surface. (Remember that car with its sticky windshield after it was parked under trees.) This sugar is now lost to the fungi that live in a symbiotic relationship with the trees and provide services to their roots: if much is squandered above ground, little ends up below ground. Poorly provisioned fungi produce fewer fruiting bodies, which snails and insects rely on for food. Little wonder that it is almost impossible for scientists to evaluate the overall balance.

It is easier to lay out the stark changes caused by commercial forest practices. By removing the original forests—that is to say, by planting monocultures of trees to harvest for timber—not only are native species displaced (in Central Europe, that would be the beech) but so are the communities of life that depend on them. Earlier, we were talking about individual cogs in the mechanism of nature, but here we are talking about the mechanism itself being replaced. Whether the new clock will work as well as the old one is open to question.

Unfortunately, the public health patrol does not concern itself with how the clock functions as a whole but only with a few forest disruptors. We've already met some of them: pine-tree lappets, pine beauties, and bark beetles. Now let's take a closer look at the latter.

6

Is the Bark Beetle All Bad?

TYPOGRAPHER, ENGRAVER—BEHIND THE delightful names hide insects that top the list of the most-feared troublemakers in our forests. They are bark beetles, and you've surely heard tell of them. The name has such a negative connotation these days that I'm frequently asked whether all the dead wood in our forest preserve isn't a breeding ground for such pests and whether it would be best to get rid of it. But Engraver Beetles & Co. pose no threat to healthy forests and happen to be quite wonderful creatures. I invite you to take a look at them in their natural habitat.

As their name suggests, bark beetles live in forests. They live in trees, but not just any old trees. Every species of beetle has its own preferred species. The spruce engraver beetle, for example, which is quite a large beetle, specializes in spruce trees and is therefore tied to places where spruce trees grow. In spring, when the thermometer climbs toward 70 degrees Fahrenheit, adult beetles come out from their hiding places under the bark where they have spent the winter and begin their dispersal flight to find a mate. But things are not quite that simple. The little males have to make elaborate preparations if they want to get lucky.

First, they target weakened spruce trees. Like all trees, spruce can defend themselves against insect attack, and who wants to die just before their first opportunity to have sex? Therefore, the beetles narrow their search to trees broadcasting scent signals that betray their weakness, because trees let each other know when they are stressed. For example, if there's a dry spell and a dangerous lack of water in the ground, the first trees to notice can forewarn all their fellow trees in the area. These can then preemptively dial down their water use to make the remaining supplies in the root zone last longer. Unfortunately, the trees' enemies also pick up on the fact that someone is in danger of running dry. Normally, spruce defend themselves against insects trying to bore into them by pushing out blobs of pitch to drown them. (Pitching them out, as it were.) If the trees are short on water or weakened in some other way, they don't have enough energy to produce pitch.

Once the male engraver beetle has found a likely candidate, it immediately begins boring into it. "All or nothing" is the beetle's motto, and if the little male is lucky, what comes out of the tunnel it drills is—nothing. It continues to make its way beneath the bark, excavating a tunnel that runs parallel to the bark fibers. It advances, fraction of an inch by fraction of an inch, backing out to expel the sawdust it creates.

This brownish frass is a red flag for foresters, because it is a clear indication that the spruce tree can no longer defend itself and is doomed. Once the beetle has succeeded in getting that far, it sends out a scent signal to summon more of its colleagues. It might sound odd to broadcast an invitation to other males at mating time, but there's method behind this madness. A brief shower might be enough to re-energize the tree sufficiently that it could quickly dispatch the courageous pioneer with a fresh portion of pitch. And so the spruce must be weakened quickly until there's no way it can recover. The more insects that bore into it, the more surely they can extinguish the tree's life force.

After a while, though, you can have too much of a good thing. If too many other males arrive, there will be enough room to build egg chambers but not to accommodate the larvae that will later eat their way out of these chambers and through the bark, forming a starburst of tunnels as they go. The result would be many starving young bark beetles. Once enough males have assembled, they send out a signal to indicate that the tree is fully occupied to keep further rivals away. The late arrivals are not left out in the cold, however, as there are usually more spruce trees in the neighborhood that can now be tested. And the odds are relatively good that these too are weakened—at least in Central Europe. After all, spruce are not native here, and they are always growing in conditions that are too warm and dry for them. Sometimes bark beetles arrive in such large numbers that they overwhelm even healthy trees. When whole stands of trees are affected, we talk of "beetle hot spots." The reddish crowns of the dying trees draw attention to themselves from a distance.

Talking of drawing attention, chemical communications have the disadvantage that enemies can "eavesdrop." For example, there is the red-bellied clerid, or ant beetle, which really does look like a large wood ant. It hunts down bark beetles and eavesdrops ever more avidly when it finds itself near them, and it's not just the adult ant beetles that eat the bark beetles in both their larval and mature phases; their larvae dig in as well. Too much chatting is a disadvantage for bark beetles, just as it is for trees.

As it calls for reinforcements (or turns them away), the little male bark beetle keeps its eye on the prize: finding a mate. It excavates a nuptial chamber under the bark and, using a different scent signal, summons female clientele. Once they arrive, there's sex and then work—mostly women's work because there's from one to three females per male. The female beetles construct more tunnels with tiny alcoves for the eggs, which are laid one after the other once construction is complete. Meanwhile, the females continue to mate to ensure they get enough sperm to fertilize thirty to sixty eggs. The bark beetle male doesn't stand idly by. Like a real gentleman, he helps by removing the frass.

Later, when the larvae have been left to hatch alone, they can eat their fill of the nutritious layers just under the bark and fatten up nicely. In the off-season, you can get a good view of their handiwork on old bark that has fallen off a tree. The farther from the egg chamber, the wider the passages chewed out by the larvae. The increasing width of the passages reflects the increasing girth of the young bark beetles. At the end of each passageway, you will find a hole. This is where the beetle emerged from its chrysalis and flew away—but not before fortifying itself with a final helping of bark. You can see the exit hole clearly if you hold the section of bark up to the light.

Development from egg to beetle takes about ten weeks, which means that it's possible to have many generations a year—depending on the weather. Cool, wet summers are hard on spruce engraver beetles, because not only are the trees better able to defend themselves then, but it's also easier for fungal infections and other diseases to spread through the insect population. (They dislike long rainy spells as much as we do.)

Fungi, however, are not always a bad thing for beetles. Some species of beetle even need damp wood where these guys have taken up residence. Consider, for example, the conifer ambrosia beetle. It avails itself of timber that is just beginning to dry out slightly. Wood in this state is the perfect place for some species of fungi to settle, because they can't grow in the wet wood of healthy living trees, nor can they grow in the dry wood of long-dead trees. The conifer ambrosia beetle leaves nothing to chance. It carries spores of bay bolete on its body and infects the wood with this fungus while it's constructing its tunnels.

The conifer ambrosia beetle goes a layer deeper than the spruce engraver beetle and makes itself at home in the sapwood, which is the living outer ring of a tree—until just a while ago, anyway. It's moister here than farther inside the tree, which means the fungi brought along for the ride can spread easily. The beetles construct a system of passageways with short ladderlike side spurs. The fungi begin to grow on all the interior walls, and the beetles and larvae use it as a source of food. The fungi blacken the wood around the passageways. The combination of blackened wood and holes lowers the value of affected timber—at least as far as forest owners and sawmills are concerned. You can easily distinguish between an attack by these guys and an attack by spruce engraver beetles, because the frass outside the bark is not dark brown but almost white. (It comes exclusively from light wood, after all.)

Holes in the trunk, stains from fungi. Clearly, bark beetles are seen as pests. And it's not just that the timber itself is worth less. In warm, dry years, the beetles can multiply so much that trees all along mountain ridges die, as you can see in the Bavarian Forest National Park.

Destruction caused by the mountain pine beetle, however, is on another scale altogether. This beetle lives in pine forests in western North America, where it is particularly partial to lodgepole pines. It behaves much like the spruce engraver beetle, except it is the females that lead the attack and summon the males with seductive scents. To shut down a tree's defenses (its flow of pitch), the beetle carries a fungus that attacks and paralyzes the living layers of bark. That way, not only are the tree's defensive mechanisms shut down, but it also cannot feed itself, and the defenseless victim can easily be colonized.

In recent years, there have been increasing reports that these beetles have multiplied so much that they also decimate healthy forests. They have destroyed approximately 55 percent of all the commercially harvestable pine in British Columbia, and enormous areas have been stripped of all their old trees.[1]

You have to wonder how this can happen. Usually, a species doesn't destroy its natural habitat. Scientists suggest it has to do with climate change. Higher winter temperatures allow more eggs and larvae to survive, and the beetles to extend their range farther north. Warming also weakens trees so that they have less energy to defend themselves against their attackers.

This certainly is part of the problem, but most studies don't mention the other part: extensively annihilating ancient forests and replacing them with gigantic monocultures that favor extreme increases in beetle populations. Moreover, rare natural fires—started by lightning, for example—have been suppressed, which means that there are many more pines growing in the forest than there used to be. And because these forests are now plantations that stress trees, there are many more weak pines, which makes it easier for the mountain pine beetles to proliferate.

The mountain pine beetle, in the meantime, is moving farther and farther north, and higher up mountain slopes. In other words, it is moving to cooler places—or to places where it used to be cooler. Here, it encounters species of pine that have never met mountain pine beetles before and therefore are not very good at defending themselves. The pine beetle's original victim, the lodgepole pine, usually doesn't give in without a fight. When a beetle bores into a lodgepole pine, the first thing the tree tries to do is pump generous quantities of pitch to the wound site. That way it either drowns the beetle or, at the very least, flushes it out. Burly beetles, however, wade through the sticky substance, and as they do, they change the chemical composition of the pitch, turning it into an invitation to others of their kind to come along and start munching their way into the tree as well.

Once the bark beetle has overcome the first obstacle, it arrives at the living cells of wood. These immediately commit suicide, releasing a potent insect toxin.[2] If there's just a lone beetle, it's killed; however, if it has been joined by colleagues summoned by the chemical call for reinforcement, the beetles will weaken the tree until, exhausted, it quickly surrenders.

There are similar wide-ranging forest collapses in Germany. In the Bavarian Forest National Park, large areas of spruce originally planted as commercial forests were put under protection, along with other trees. Once foresters could no longer cut down trees that had been attacked or pump them full of chemicals, the engraver beetle ran riot, just like its cousin in North America—with identical results. Mountain ridges were covered with dead trees that had been attacked. Hikers were shocked to find nothing but a bleak landscape haunted by cadavers instead of the green paradise they had been expecting.

And now we should ask ourselves once again whether bark beetles really are pests. As far as I'm concerned, the answer is a resounding no. These insects prey on the weak, so they can only damage trees that are already in trouble. The mass reproduction events that allow the beetles to overcome healthy trees happen only when people have changed the natural rules so much that the beetles can gain the upper hand. This could be through creating plantations or emitting the pollutants that lead to climate change. Ultimately, we, not the beetles, are to blame for upsetting the carefully calibrated balance of nature. Instead of blaming the beetles, you could see them as an indicator that things are not as they should be. You could argue that all they are doing is exacerbating a situation that is already out of balance, making it all the more urgent that we change course to bring us more in line with the natural order.

The coniferous plantations in Central Europe—vulnerable artificial arrangements of nonnative trees—could gradually be replaced by native deciduous forests. There are bark beetles adapted to attack these trees, as well; however, since Beeches, Oaks & Co. are much more at home here than spruce or pine, they usually have no problem repelling bark beetle attacks. To label bark beetles as pests diverts attention from the real causes of the problem. And individual trees that are attacked because they are already weak are a vital source of food for ant beetles, woodpeckers, and many other species. You could say that bark beetles open the door for creatures that make their living off dead wood. By multiplying in former plantations, they create a temporary paradise for detrivores. And in the ravaged stands of spruce in Germany's national parks, the next generation of trees is already primed and ready to go. A whole generation of trees, including many that are deciduous, is standing by to form a solid foundation for the old-growth forests of the future, which means that bark beetles are more than just funeral directors; they are midwives, as well.

It's somewhat easier to see what's going on in the case of large dead animals. Dead animals? Yes, a dead animal is an ecosystem unto itself, a bit like a small planet in the universe of nature. A planet that is a bit on the smelly side, perhaps, but one that has been given far too little attention until now.

7

The Funeral Feast

So far, we've passed over a particularly tasty treat for many species: the carcasses of large mammals. Fascinating things happen around these dead bodies. Do you find that disgusting? That's understandable; however, strictly speaking, we're surrounded by the dead bodies of animals all the time, and unless we are vegetarians, we interact with them (albeit briefly) almost daily—on our plates. The main difference between our meals and the many dead wild boar, roe deer, and red deer out in the wild is that the process of decomposition has barely started, which allows us to enjoy our meals safely.

Many animals tolerate or even require various stages of putrefaction in their food and are perfectly happy devouring servings of meat that stink to high heaven as far as we're concerned. And there are a lot of these servings of meat to be consumed. Every year in Central Europe alone, millions of roe deer, red deer, and wild boar die a violent death. And although in Germany, for instance, a lot of wild game is shot (about 1.8 million of the three aforementioned species, according to the German Hunting Association), many more die a natural death. What happens to their bodies? Off the top of your head, you'd say: they decay. That is to say, they rot and eventually, after smelling awful for a while, they become humus. But who facilitates this process?

Let's start with the larger facilitators: bears. They have extremely sensitive noses and can smell a side of meat from many miles away. Together with other large predators, such as wolves, they can consume most of the meat off a dead animal within a few days. What they can't consume, they bury, so they have a hidden supply of food.

Birds are early responders, as well. Whereas vultures circle over fresh carcasses in the African savanna, noisily laying claim to them, in northern latitudes, ravens take their place. Ravens are the vultures of the North, and they, too, patrol their territory from above to see where a deer or wild boar might have met its end.

Dead animals are often the cause of fights, and wolves lose out when brown bears turn up. Then it's best for the pack to head for the hills, particularly if they have pups, which a bruin could easily scarf down as a snack. Ravens have a role to play here: they spot bears from afar and help wolves by alerting the pack to approaching danger. In return, wolves allow ravens to help themselves to a share of the booty—something the birds wouldn't be able to do without the wolves' permission. Wolves would have no difficulty making a meal of ravens, but they teach their offspring that these birds are their friends. Wolf pups have been observed playing with their black companions; the young wolves imprint on the smell of the ravens and come to regard the birds as members of their community.

Wolves and ravens might live peaceably with one another, but other species fight over food resources. Apart from the black birds, there are other feathered parties, such as bald eagles or kites that would love to haul away a portion of the booty. With all the commotion and clamor as animals wait their turn, the ground around the carcass gets torn up. The plants are shuffled, because seeds that would otherwise have been stifled in the matted grass now have their moment in the sun. Things also change for vegetation that is not disturbed. Rotting flesh serves as fertilizer—for the plants, deer carcasses are simply overgrown salmon. The more robust growth and greener color of the grass and nonwoody plants for about 3 feet around the carcass are evidence of the nutrient boost.[1]

And what happens with all the bones? After the flesh has been eaten or has rotted away, there should be huge numbers of bones lying around in field and forest, bleaching in the sun. But no, even I on my daily rounds as a forester have never come across a dead animal's final resting place, and only very occasionally do I come across a skull.

There are two forces at work here. Sick or weak animals separate themselves from others of their kind to hide in the undergrowth, or on hot summer days they wander near or into small streams to cool any wounds they might have. Here, they wait for death. That makes sense, because this way they don't endanger their kin—weak animals attract the attention of predators. Also, in a secluded spot there's no one to disturb them in their final hours. Usually, it is our sense of smell that leads us to dead animals in places like this; bones lie quietly hidden from sight under bushes. Because bones basically don't break down, and because every now and then animals certainly die away from the protective cover of vegetation, over time you'd expect to find bones scattered all over the place. But that is not the case, because there are plenty of takers for the dead animals' final remains.

There are mice, for example. They seem to love bones, and they gnaw away at them until there's nothing left. Calcium and other minerals are mostly what they're after; bones are for mice what salt licks are for cattle (or salted pretzels are for us). If the bones are still fresh, bears eagerly crack them open for the fatty marrow inside—a delicacy that no one will fight the bruins for, not even wolves. Although some dogs like to chew on bones, the gray-coated hunters clearly don't think much of this tedious detail-oriented task, but it is an important one, especially for other species. Just how important this task is becomes clear everywhere bears have been eradicated, including in Germany, because it is only when the hard outer coating has been cracked that daintier creatures get their turn.

Take the bone skipper, which disappeared without a trace until it was rediscovered in 2009.[2] This bizarre insect with its tiny orangey-red head looks like a creature from a fantasy world, and it doesn't behave like other flies, either. The bone skipper likes it nice and cold. It is out and about mostly on winter nights, on the lookout for dead animals and cracked bones. Here, it feasts and lays its eggs. By the nineteenth century, however, there were no longer any carcasses out in the open in Central Europe—thanks to stricter rules about hygiene. At the same time, bears were driven out, and so things became grim for the bone skippers, and in 1840, they were declared extinct. In 2009, however, the Spanish photographer Juli Verdú took a picture of what he thought must be a fly that had flown in from the tropics. Researchers at the University of Madrid recognized it as the long-lost insect, which could then be crossed off the list of extinct animals.[3]

We spoke earlier of ravens as the vultures of the North, but we should also mention vultures themselves. Griffon vultures searching for dead animals regularly fly over Germany. On the website Club300, amateur ornithologists report sightings of these unusual visitors every year.[4] If there were something for them to eat, several of them might once again make their home here, but as it is, all they do is make flying visits that pass unnoticed by most of us. Griffon vultures, like bone skippers, have been declared locally extinct in many places in the world.

UNTIL NOW, WE'VE been concentrating on the carcasses of large animals. These are usually fastidiously disposed of, but below a certain size, this cleanup no longer happens. There are numerous remains of small mammals out there, and they vastly outnumber the large ones. Take mice, for example. Up to 250,000 of these little rodents scurry around per square mile, living on average just four and a half months. Young mice are sexually mature by two weeks, and after another two weeks, about ten babies are born.

Let's assume that during one growing season, there are five generations of ten mice each for every mouse pair. In particularly fruitful years, that would mean the 250,000 animals (or 125,000 pairs) per square mile would lead to 6.5 million mice scampering around—not all at the same time, of course, because most of them would have died from disease or been eaten as the season progressed. Over the course of the season, then, there could be up to 6.5 million dead mice lying around. If each weighs an average of 1 ounce, the total weight of these dead bodies would be 200 tons, the same as about twelve thousand roe deer. That's way too much mouse meat to be carried off by buzzards, foxes, or cats, which leaves plenty for others to exploit.

One of these others is a pretty black-and-orange striped beetle appropriately called the sexton or burying beetle. I encounter it frequently on my walks through the forest—it's so striking that it's hard to miss. Even though the adults hunt insects, they can't resist the delectable scent of fresh carrion.

For sexton beetles, a mouse carcass is attractive not only as a hearty meal but also as a good place for their offspring to get their start in life. It's often the males who first occupy the prize. They triumphantly raise their rear ends and discharge a scent message to attract females. Their goal is clear: sex. Rivals, however, also get the message and fly on over. Fierce struggles ensue, and the losing beetle has to beat a hasty retreat. When a female turns up, the work begins.

The beetles dig tirelessly underneath the mouse, dragging it down by its fur. In the process, a lot of the fur gets bitten off and the carcass gets coated with generous quantities of saliva. That doesn't sound very appetizing, but it makes the mouse more slippery. And so the dead animal gradually descends farther down into the soil, until it eventually disappears completely—safely beyond reach of other carrion eaters.

The beetles take frequent breaks to have sex. After all the work is over, the mouse doesn't look much like a mouse anymore. All that pulling and pushing has transformed the carcass into an elongated pellet. The female now lays her eggs alongside it. Unlike many other insect parents, sexton beetles stick around after their larvae hatch. The youngsters' mouthparts are not strong enough to chew meat, and so the mother feeds her little ones, which lift their heads and beg for food like baby birds in a nest.

As researchers at Ulm University discovered, something else happens to the beetle mother: she loses interest in mating. Not only that, even if the male were to get lucky, it wouldn't do any good, because his beloved is now completely infertile—at least as long as she still has her full complement of babies. As soon as a couple of the little ones go missing (perhaps because they died or were eaten by some animal), her desire for sex returns. The male immediately gets wind of the change and goes berserk. The scientists observed up to three hundred copulations—more than when the male initially laid claim to the carcass. The female quickly lays new eggs to replace her loss. If, in this flurry of activity, she ends up with too many babies, she soon fixes things by killing the extras.[5]

If neither bears nor wolves take care of a carcass (or, in the case of smaller carrion, sexton beetles), then smaller creatures take over. Heading up this squad of cleaner-uppers are blowflies. In Germany alone, there are more than forty species that are magically attracted to the smell of dead bodies. The meat shouldn't be so far gone that it stinks, because these insects prefer to descend on fresh food. For example, if you leave a portion of barbecued meat on your plate in summer, it often takes only a few minutes for the first flies to appear.

I found out for myself just how fresh these iridescent blue flies like their meat. On a hot summer day years ago, I came across a roe deer that had lain down in the undergrowth. It was badly injured and had a large wound on its hindquarters. There were already hundreds of fat white maggots crawling around in it—the children of blowflies. With a heavy heart, I put the deer out of its misery. Some species, such as toadflies, even attack perfectly healthy animals. They lay their eggs on the skin of toads and when the larvae hatch, they crawl up into the toad's nostrils, where they begin to eat their host's head from the inside out. The toad briefly shuffles around zombielike before finally giving up the ghost.

Usually, however, blowflies are the first guests to arrive at a fresh carcass. Hundreds of flies lay thousands of eggs, preferably in exposed places such as the eyes. The fast-developing larvae spread from there to the whole carcass and cover it so completely that other insects have little chance of finding an empty spot to lay their eggs. The bone skippers are the last to appear, content as they are with the leftovers—which is to say: the bones.

There's a simple way—in Germany, at least—to support bone skippers and numerous other species that depend on very large carcasses. We could let dead deer and boar lie, at least in national parks. Usually, people hunt in these parks, and the carcasses of wild animals are taken away by foresters. But, in national parks, anyway, natural processes are supposed to be able to play out, and the carcasses of animals are part of this cycle. We won't be able to see the red-headed bone skipper for ourselves, because it's mostly active on cold nights. Even so, it's good to know that this ecosystem, with its sometimes bizarre-looking creatures, has another chance at survival.

Talking of the night, there are other representatives of the insect realm that also like the dark and even light small lamps of their own. These lamps light the way to love, intrigue, and sometimes even gruesome death.

8

Bring Up the Lights!

LIGHT IS OF utmost importance in nature. Ultimately, almost every creature on this planet lives off processed solar energy. Photosynthesis produces sugar, which fuels plant life and therefore, indirectly, human and animal life, as well. In the natural world, there's clearly a struggle for every ray of sunlight, for every smidgen of energy. Trees demonstrate this particularly well. The only reason they grow as tall as they do is to rise above the other plants and bushes that compete with them for light.

Developing mighty trunks and crowns takes a great deal of energy. For example, a mature beech contains up to 14 tons of wood, which, if burned, would release about 42 million kilocalories of energy. To put this in perspective, a person—depending on how active they are—burns between 2,500 and 3,000 calories of food a day. (Food calories are actually kilocalories, even though we refer to them simply as calories.) This means that a mature beech stores enough solar energy to feed a person for forty years—if the human gut were able to digest wood. It's no surprise that it takes decades to produce this much wood, and that's why trees have to live to be so old. A forest ecosystem, then, is basically an enormous storehouse of energy.

So far, so good, but light is important for completely different reasons, as well. Its energized waves stimulate the retina at the back of the eye, where they are transformed into information. Most animals have developed their vision to interpret light, which means, of course, that some light needs to be available. Apart from the fact that the enormous crowns of trees block up to 97 percent of the light in a forest, there's yet another problem for animals that need light waves in order to see. Half the time—at night—there's precious little light around. Only the faint glow of the stars, augmented by the brighter light of the moon, offers some relief from the darkness. But when it's cloudy, as it so often is, it can get pitch dark out there. And so why not make a virtue out of necessity?

Although the title of this chapter is "Bring Up the Lights!," for some plants and animals "Turn Down the Lights!" would be a better call. Plants and animals are nocturnal for many different reasons. Some flowers bloom only when it's dark, because they want to avoid competition. During the day, a multitude of nonwoody plants, bushes, and trees go to great lengths to stand out from the crowd, all vying for the attention of pollinating insects. Let's consider honeybees, which are important pollinators. They are limited in how many flowers they can visit, and if the floral host is too large, then many flowers miss out on pollination and don't form seeds. To avoid that fate, plants make use of the full range of colors in nature's palette. In addition, they send out sweetly scented invitations. What smells good to us also smells good to insects, because sweet smells indicate where delicious nectar is to be found.

Some plants opt out of the colorful daytime chorus of visual and olfactory communication and shift their bloom time to the hours of darkness. Their names—evening primrose or moonflower—often point to their nonconformity. After the sun sets, most other flowers shut down, so you could say that the competition goes to sleep. Now the insects can focus all their attention on the few remaining plants offering nectar. It's too bad that the honeybees, like most of the flowers, also stop and take a break. They returned to the hive a while ago to spend the night processing their haul and preserving it as honey.

But there are insects that work the night shift. Take moths, for example. I'll come right out and say it. Even though I am an animal lover, moths are not high on my list, but in my defense, there is a story here. Years ago, when we came back from a family vacation in Sweden, we noticed when we finally collapsed onto the couch after unpacking the car that there were tiny moths flying all around us. A nasty feeling came over me, and I lifted a corner of our wool rug. Horror of horrors! Thousands of larvae were wriggling in the wool, and a blizzard of disturbed moths flew up and fluttered around the living room. We quickly rolled up the rug and banished it to the garage. The incident made me feel queasy around moths, and a flicker of queasiness returns anytime I come into contact with wool.

Moths and butterflies belong to the same order (Lepidoptera), but there are distinct differences between them. One is that butterflies fly during the day, and most moths fly at night. Another is that butterflies are colorful, whereas moths tend to be rather drab, but there is a good reason for that. Butterflies use their colorful patterns to communicate information to other butterflies and enemies, but moths have a completely different strategy. If they are to survive, moths need to remain as unobtrusive as possible and blend in with their surroundings as best they can, because these little winged creatures spend the day somewhere on the bark of a tree, where they have to avoid the attention of birds that might eat them.

Their avian attackers sleep at night, which puts moths at a distinct advantage when they visit the sweet chalices of night-blooming plants. How lucky for moths that most birds agree with the plants; the hours of darkness are not to their liking and they avoid them. But because this interplay between species has been going on for millions of years, it should come as no surprise that predators are standing by to exploit the situation.

These predators are bats, which hunt moths in the warmer months of the year. And because light is in short supply at night, bats use ultrasound to locate their prey. I think it's entirely possible that bats use their calls and the sound waves reflected back from objects to help them construct images inside their heads—in other words, they "see" objects using sound.

Scientists assume that thanks to the echoes bats receive, these nocturnal hunters have a very clear idea of whom or what they are dealing with. A leaf falling from a tree creates a different ultrasound pattern from the flap of a moth's wing. Bats can detect wires no more than two one-thousandths of an inch thick, and it's possible that they "see" their surroundings in much more detail than we do with our eyes during daylight hours.[1] After all, when we see things, we're doing nothing more than interpreting waves reflected off objects. The only difference is that we're adapted to light instead of sound—that, and the fact that bats have to shout all the time if they want to see something.

This shouting is not long and drawn out, as it is when we want to elicit an echo while we're out hiking in the mountains, for example. Unlike us, these nocturnal hunters make a series of short, rapid calls—about a hundred a second. And with bats, it's all about volume: the sounds they make can be as loud as 130 decibels, which would hurt our ears if we could hear them (but ultrasounds are out of our range). Unlike lower frequency sounds, ultrasounds are quickly swallowed up in the air, and after they've traveled about 300 feet, there's not much left to hear. Even so, on summer nights it can get quite noisy out in forests and fields—in the upper frequencies, at least.

To camouflage yourself from the reflection of light waves, or, to put it more simply, from being seen, all you need is colors and patterns that blend in with your surroundings. The same is true if you want to camouflage yourself from sound waves. In this case, blending in with the surroundings means that, as a moth, you reflect back as little sound as possible.

You can test out how this works the next time you hike in the mountains.

Your calls to elicit echoes are reflected particularly clearly if the slopes around you are not covered in trees. If the slopes are covered in dense stands of trees, you usually won't get an "answer," because the trunks and crowns swallow sound. To exploit this effect for their own ends, moths grow a mini forest. Their bodies look furry, and these "hairs" ensure that the sound waves are not reflected back crisply and clearly but are deflected in different directions so that the bats can't get a clear picture of the moth. Unfortunately, this deflection has its limits, so these insects need more tricks to increase their chances of survival.

There's a regular arms race going on between moths and bats, and at least some of the moths are catching up. Over time, some moths have evolved to hear sounds at extremely high frequencies. The highest frequencies bats use when they hunt are around 212 kilohertz. In contrast, human hearing breaks down at frequencies higher than 20 kilohertz.

Although most moths can hear higher frequencies than we can, many cannot match the frequencies of bats. The result is that these moths can't hear the bats coming (bat wings make hardly any noise at all), and they are taken completely by surprise when the bats attack. But that's not the case for every species, as reported by one of Hannah Moir's research teams at the University of Leeds. Greater wax moths can hear sounds in the range of 300 kilohertz—the highest hearing score in the animal kingdom—even though the greater wax moth's ear has a really simple design: it is made up of a membrane with just four receptor cells. (In comparison, there are twenty thousand so-called hair cells that vibrate in a human ear, in addition to other structures that convert sound into neural signals.)

As Moir and her colleagues report, the moth's hearing is way more sensitive than it needs to be. If bats don't produce sounds much higher than 200 kilohertz, why do moths need to be able to hear higher frequencies? Especially as bats are unlikely to upgrade their calls: sounds in frequencies higher than the ones the bats are currently using are quickly muffled by the air around them and are therefore not helpful when the bats are trying to create echoes.

Why, then, have greater wax moths developed this extraordinary ability? The researchers speculate that the moths probably have something completely different in mind. They, too, communicate at high frequencies, for example, to find a mate. Their closely spaced courtship calls are within the range of the bats' more widely spaced echolocation calls. The simple construction of their ear means that greater wax moths can distinguish closely spaced signals better and faster—six times faster—than other moths. (Some individuals, it turns out, have more sensitive hearing than others.) Thus, the moths can flirt in peace, because they can hear the echolocation calls of their greatest enemy loud and clear over their own calls, and they can make themselves scarce if they need to.[2]

The greater wax moth is not the only species that has armed itself against bats. Some moths interfere with bat echolocation systems by producing decoy calls—clicks in the ultrasound frequency that confuse approaching bats. The moths basically disappear in static on the bats' radar. The great tiger moth—a member of the subfamily Arctiinae, many of which have hairy caterpillars commonly known as woolly bears—produces such a fearful din that alarmed bats call off the chase.

How do moths escape once they've heard their enemy? Bats fly much faster than moths, and they're also more maneuverable. Therefore, there is just one basic defensive response when danger is closing in: as soon as the moths capable of hearing ultrasonic calls register the sounds of a search, they fall to the ground in terror. The bats have little chance of relocating their prey in the grass. Despite the moths' defensive tactics, however, bats eat their fill at night. There are always careless moths to be caught, and there are mosquitoes, as well. So the bats catch up to half their own body weight in insects a night. (On a night of dining on nothing but mosquitoes, that would be about four thousand mosquitoes per bat.)

HUNTERS AND PREY coexist in a delicate balance that gives each a chance to survive. But artificial light can upset the scales. There is, of course, just one relevant source of natural light at night: the moon. When it shines, animals use it to orient themselves. It serves as a kind of compass. When moths fly at night and want to fly in a straight line, they are careful to keep the celestial body at a set angle to their flight path. That works wonderfully well—until an artificial light crosses the tiny flier's path.

The insect assumes this light must be the moon. Confused, it tries to fly so that the moon is always on the correct side—for example, to its left. With the moon, that's not a problem because it's almost infinitely far away. With the light, which is close by, the insect quickly flies past. The source of the light is now behind it. It keeps correcting course, and its course corrections lead it to fly in an ever-diminishing circle. Eventually, the moth crashes right into the light itself. It keeps starting again to try to get away—and every new attempt at escape ends in failure.

Some moths die of exhaustion; others meet their end more quickly. Over time, many bats have specialized in patrolling streetlamps. They can eat their fill of insects more quickly here, because all they have to do is check out one lamp and then the next to see if yet another moth has become confused by the artificial moons. Even house windows lit up at night can set the stage for similar small dramas, as my wife and I have observed. As we sit comfortably on the couch watching a film in the evening, moths gather at our living room window. Every now and again, shadowy bats flit by—until, finally, all the moths disappear.

There's a whole host of other insects that become confused by artificial light in a similar way. Like moths, they are magically attracted to yard lights that are supposedly lighting up their surroundings in an environmentally friendly manner. There's usually a solar cell sitting on top, which is great because it means the power source is ecologically correct, and that means there's no issue with leaving the light on all night long. This delights the large number of spiders that profit by spinning their webs here. If the situation persists over time, the tiny ecosystem around the light changes, because some species disappear entirely (into the spiders' stomachs). If it were just one light, that wouldn't matter so much, but it does matter when there are thousands upon thousands of them—as there are in urban areas.

There have, however, been additional sources of light in the landscape since long before people started lighting things up. On warm summer nights, thousands of tiny greenish lights glimmer along the edges of forests and shrubby areas. These are fireflies (sometimes called lightning bugs), which strut their stuff when it gets dark. The brightness of their light may be a thousand times less than that of a lit candle, but the efficiency with which they turn energy into light is exceptional. Using our latest technological advances, we can convert 85 percent of energy into light; fireflies can convert 95 percent. They need to be frugal because as adults, they don't eat anything—at least not in most cases (there are gruesome exceptions, but more on those later).

Fireflies really should glow red, because the purpose of their nocturnal light show is love. In the most common species in Germany, *Lamprohiza splendidula*, the females light their lamps on the ground. Fireflies are also referred to as glowworms, even though the light you see at night comes from the adult beetles; however, the females have stunted stubs instead of wings and cannot fly. With their pale yellow abdomens dotted with tiny lights, they really do look like luminescent worms.

These earthbound females don't turn on their lights until they spy a male above them. Males can fly, and they scour the neighborhood in search of a mate. Their last two body segments are protected by a transparent chitin casing that allows them to shine their light downward. That way they don't reveal their presence to enemies flying above them, while at the same time signaling, "Look what a great guy I am" to the females below.

When one of the solicited females gets the message, she turns on her light as well, inviting the Casanova to land, which he does right away. There's mating and then egg laying. The larvae that hatch eat a great deal. They are partial to snails and will tackle specimens up to fifteen times their own weight.[3] They kill the snails with a single bite, and then eat them at their leisure. The firefly larvae keep eating until they're ready to burst, and when their stomachs are full, it's time to take a nap. How long they pause to digest depends on the size of their meal, but their postprandial nap could last for days.

Depending on the species, it can take about three years until the offspring develop into sexually mature beetles. Given how long they spend in their larval stage, the worm part of the common name glowworm fits them well. The luminescent beetles live for only a few days: the male dies soon after mating, and the female dies right after she has laid her eggs. And so their glow is literally a final flicker of a life that ends on an ecstatic high. At least it does when all goes according to plan. Unfortunately, there are always disruptive elements lurking out there in nature.

The glowworm's peaceful illumination with love in mind is abused by others for their own ends. In New Zealand and Australia, there are gnats in the genus *Arachnocampa* whose larvae also light up. These antipodean glowworms live in caves and deep in the rain forest, where they gather in groups high up on the ceiling or in the canopy. The larvae need a dark, humid place where the air is still—perfect conditions that only caves and dense forests can provide. The larvae spin sticky threads covered in tiny drops of moisture that light up.[4] The effect is magical, and caves where these larvae live have become popular tourist attractions. But wealthy tourists aren't the only ones attracted to the lights. Insects also come, probably because they confuse the twinkling droplets with stars in the night sky. Thinking they are flying out in the open, they get caught in the sticky threads and end up in the stomachs of the hungry gnat larvae. Researchers have discovered that the hungrier the larvae are, the more brightly they glow.

A North American beetle in the genus *Photuris* employs an even more deceitful tactic. Fireflies have developed a variety of techniques using light to draw attention to themselves. After all, there are different species and if all each of them did was light up, they could easily get confused when it came time to find a mate. Therefore, there's a kind of Morse code out there—flashes with a particular rhythm and frequency that attract beetles of the same species. Human Morse code would be too primitive for the beetles: on/off and long/short just don't contain enough information. Using up to forty flashes a second with varying degrees of brightness, the beetles are capable of a much wider range of signals.[5] And so cheery winks are a way to find the love of your short life—unless you belong to the genus *Photuris*.

Females in this genus imitate the light signals of a different species to lure their males, which flock eagerly to them.

When the males land, instead of an amorous adventure, they find the eager mandibles of *Photuris* females. They need the males not only for their calories but also for the toxins in their bodies. These then protect the females from being eaten by spiders, which also notice their light signals and, in the absence of toxins, would be happy to accept the illuminated invitation to dinner.[6]

It's not just insects that use light as an attractant. The deep-sea anglerfish possesses, as its name implies, a fishing rod. The rod sits on the top of its head, dangling a bioluminescent lure out in front of its mouth—a mouth that bristles with needle-thin, dagger-sharp teeth. The light source attracts other fish like a magic wand, and you can imagine how their visits end. Humans get similar results when they use lights to fish. Japan, for example, uses this technique on a massive scale.

Light is tremendously attractive whether you're on land or in the water. And this brings us back to the problem of humans lighting up the night. It is startling when you look at a satellite map of the earth at night and see how much of its surface is lit up with artificial light. You can easily gauge for yourself how severe the impact of artificial light is where you live by stepping outside your front door at night. Can you see the Milky Way on a clear night? If you don't even know what the Milky Way looks like, there are certainly too many artificial lights in your neighborhood, because there's no way you can miss this band of light if conditions are favorable.

Visibility is further diminished by air pollution, which diffuses light particles, so the number of stars visible to the naked eye can decrease from about three thousand to fewer than fifty. And the delicate light signals from fireflies are similar to weak stars, aren't they? The more artificial light there is in this world, the more the confusion I have described happens in the animal realm, and the less successful those species are that produce light.

The confusion can be fatal. Freshly hatched sea turtles orient themselves to the glittering waves of the ocean lit up by the full moon. As soon as they've scrabbled their way out of their sandy hiding places, they set out in this direction as quickly as they can to escape hungry predators. Problems arise if the beach is close to a brightly illuminated sea walk or a strip of hotels. The tiny turtles set out toward the artificial light sources by mistake, getting farther and farther away from the safety of the water. It's no surprise that the next day many of them fall victim to gulls or die of exhaustion.

Even weather phenomena get turned upside down thanks to the glare of electric lights. Clear nights used to be particularly bright, which makes sense because the moon and the stars can shine down uninterrupted. Once our eyes have had a few minutes to adjust to the dark, we can walk around outside without any problem. Today, we can often walk around even on cloudy nights—weather that used to mean complete darkness—because clouds reflect urban lighting way out into the surrounding area, inadvertently lighting up the sky in a way that does neither people nor animals any good. Who among us likes to sleep with the light on?

And yes, you read that right. Artificial lighting has negative consequences for people, too. We have an internal clock ticking inside us that is regulated by light. The blue wavelengths of light are particularly important for us, because they determine whether we feel wide awake or tired. Our eyes contain the photopigment melanopsin. When it's hit by blue light, it signals to our brain that it's daytime. Usually, the system works really well. In the evening, at sunset, the light spectrum shifts to red and we automatically feel tired.

Problems arise when we watch television at night instead of going to bed, because the flickering images on the screen contain a great deal of blue light. It's no wonder so many people suffer from sleep disorders. Staring at the screen primes the cells in our body to be ready for action instead of for sleep. The makers of smartphones are trying to get a handle on the problem by adjusting the screen colors after a certain hour so that users feel sleepy as they surf and chat.

What about animals? How can we help creatures that can't escape all this light? You can give your fellow creatures some relief by closing your blinds or drawing your curtains at night—this is easy to do and blocks a widespread source of light. You don't need to keep your outside lights on all night, either. We have motion-detector lights along the driveway leading up to our forest lodge, and they come on briefly only when they're needed.

The greatest source of nighttime illumination, however, is streetlamps. These days, most radiate an orange-red light that reflects particularly well off clouds, making the problem of lighting up the night sky even worse. I was pretty excited for a while when the old white fluorescent tubes were replaced by these modern, energy-efficient sodium vapor lamps. I noticed that the undersides of clouds were glowing increasingly red, and there were nights when the radiance in the clouds drew my attention to Bonn, some 25 miles away, but I attributed the increasing brightness of the night sky more to the increasing sprawl of the city than to the change in lighting. And now? There's yet another change, this time to LED bulbs, which use even less energy. And if these streetlamps are more focused—that is to say if they only shine down (which is where the light is needed)—and if they are also turned off after midnight, these would be big steps in the right direction.

Although there's still a need for significant improvements at night, when the sun lights up the day, there are heartening signs of progress in environmental protection way up in the sky. In fall, impressive flocks pass by in formation on their way to interfere with the production of Spain's famous Iberian ham.

9

Sabotaging the Production of Iberian Ham

EVERY YEAR I look forward to fall, or, to put it more precisely, I look forward to the cranes. You can hear the trumpetlike calls of the migrating flocks from many miles away. And after all these years of keeping an ear out for them, I can pick up their distant calls even if my living room windows are closed. In recent decades—thanks to better environmental protections, such as the reclamation of wetlands—the number of birds has risen to such an extent that Eurasian cranes are no longer listed as endangered. Day after day, one formation after another flies over our forest lodge, and sometimes the birds fly so low that I can hear the sweep of their wings.

What drives birds to fly to distant lands as the seasons change? How do they find their way? Bird migration is a worldwide phenomenon undertaken by about 50 billion birds. Mass aerial movements are happening all the time, because somewhere in the world there's a change from summer to fall, from winter to spring, or from the rainy season to the dry season. And as seasons change, so do food resources.

Once the cold sets in here in the Eifel mountains, insects hibernate, dozing deep underground or under the bark of mighty trees. Some species even make themselves comfortable in the relative warmth of the hills constructed by red wood ants. In their chosen hidey-holes, the insects are mostly out of the reach of birds—as are most other small animals that birds prey on, and so many feathered species set out for warmer, more productive climes.

Most researchers believe that the urge to fly to other places as the seasons change lies in the birds' genes. To me, this makes the birds sound as though they are some kind of biological automatons operating in response to a preprogrammed code, incapable of making their own decisions about where and when to fly. But apparently, they do make decisions, as the Estonian scientist Kalev Sepp and his colleague Aivar Leito discovered.

Since 1999, Sepp and Leito have fitted numerous cranes from their homeland with radio collars so that they can track their migration routes. To the scientists' surprise, they discovered that over the years, the birds switched between three different routes. That is a clear argument against the route being genetically fixed. It also seems to rule out learning the route from older birds—which until then had been another theory entertained by scientists. Sepp concludes that the birds must get together and somehow discuss where they have the best chance of finding good breeding sites and food.[1] And this brings me to the subject of this chapter.

Cranes, with their assignations and meetings in particular places, really do sabotage the production of Iberian ham. Not intentionally, of course. The birds have not the slightest interest in pigs. But they are well aware that special treats await them in Spain and Portugal: acorns—particularly acorns from the holm oak woodlands in the Extremadura region in Spain, where they know they can find copious quantities of these nuts. No wonder the cranes that fly over our forest lodge have decided that this paradise is the place where they will spend the winter. Here they can build up their strength and survive the cold season well nourished. However, other inhabitants of the Extremadura also prize these treasures, and these are the local farmers who rely on the acorns to fatten their pigs.

What we're talking about here are the famous Iberian pigs from which *jamón ibérico de Bellota*, or acorn-fed pure Iberian ham, is made. Most of the pigs are raised in an environmentally friendly manner: they spend time roaming through holm oak forests, and a good portion of their diet is the nonwoody plants they find there and, above all, acorns. And that's how it used to be in Central Europe, as well. Pigs were driven into forests in fall to fatten up on acorns and beechnuts. In those days, animal fat was still prized. The term "mast years" comes from these times. Mast years are years when there is massive production of acorns and beechnuts, and they cycle around every three to five years.

Back to the Extremadura. The holm oaks were once an important part of the ancient forests that grew here. Over the course of the many thousands of years of human history, most of the forests on the Iberian Peninsula have been cleared. Different species of trees have been planted and the character of the landscape has changed. And so today, in addition to conifers, there are more and more eucalyptus plantations. Eucalyptus grow rapidly, much more rapidly than the ancestral oaks; therefore, they are good trees to grow if you want to produce more timber.

These changes have been catastrophic for native ecosystems. Eucalyptus plantations, in particular, are considered by environmentalists to be green deserts. The trees' essential oils (which taste so refreshing in throat lozenges) are responsible for an explosion in the number of forest fires. Today, we associate Southern Europe with forests fires, but that never used to be the case. Natural deciduous forests left to their own devices do not burn, and fire was not part of the ecosystem in these latitudes.

The changes make the holm oaks that remain that much more important, even if today they no longer grow naturally but need a helping hand from farmers to get started. The driving force for farmers is not only the production of timber but also the production of acorns for their pigs. And this is where the cranes enter the picture. It's not a problem for the farmers if the birds help themselves to some of the nuts. The problem is the sheer number of birds. In the past few decades, their numbers have risen dramatically, which is good news.

According to the World Wildlife Fund, in the 1960s, there were only about 600 breeding pairs left in Germany. Since then, the number has risen to more than 8,000. Over their whole range, which includes the northern parts of Europe and a large part of northern Asia, current estimates peg the population of Eurasian cranes at about 300,000 individuals. And more and more of them are heading for Spain.

Clearly, more cranes leaves less food for the pigs, which affects ham production. It's a dilemma. Keeping pigs motivates people to preserve oak forests, which, in turn, provide important winter food for cranes. If pig farming loses its luster, at least part of the motivation for preserving oak forests is lost.

Is there a way out of this dilemma? I think there is, and the solution sounds simple: more deciduous trees in Spain and Portugal would benefit all parties. It's true that oaks don't grow as rapidly as eucalyptus or pines, and they are not as easy to maintain on an industrial scale. However, they do produce timber that is sought after, and they also provide feed the farmers want for their pigs, which other plantation trees can't offer. In addition, if more oaks were grown, the danger of forest fires would be reduced considerably, and the ecosystem would become attractive to other species again. (We haven't even touched on squirrels, jays, and the thousands of other animals and plants that depend on oak forests.)

Of course, in a democracy, you can't just issue a decree to increase the size of forests, but subsidies (which I don't usually support) would be the way to go here. When I see how the factory farming of animals profits from government incentives, I can't see that it would be difficult to do something to promote the peaceful coexistence of pig farmers and cranes. After all, it's not the birds that are overtaxing the ecosystem. It's the fact that so little oak forest remains that makes the problem so acute. And what if one day there were many more holm oaks? Wouldn't that mean that the crane population would go into overdrive? No, it wouldn't. The number of cranes depends mainly on the size of the area that offers suitable breeding sites. And, unfortunately, wetlands continue to shrink in Europe, which means that the increase in numbers will eventually level off.

If we were all to dial down our demands a bit, there would be enough space for our fellow creatures. In this sense, the crane is a good ambassador for the environment, one that will, I hope, be around for a long time and in great numbers, with its noisy flight and trumpeting calls to remind us of the early days of the conservation movement.

But what are we to do until the oak forests expand? Couldn't we just feed the cranes while we're waiting? That brings up a basic question about supporting our feathered friends, and this question has little to do with science and more to do with our emotions. We feel sorry for birds in winter, don't we? Those that don't fly to warm southern climes sit freezing on the branches of bushes and trees, puffed up like fat feathery balls while we watch them through the windows of our centrally heated homes. As they, like us, are warm blooded, they have to keep their bodies warm. And, for birds, this means maintaining body temperatures between 100 and 108 degrees Fahrenheit—which is substantially warmer than ours.

Luckily, nature has equipped birds with an outfit—a snug feather coat—that helps them keep warm. There's a reason we stuff our winter jackets with down: it's an excellent insulator. When birds fluff out their feathers, they trap a thick cushion of air, and when they puff out like this, their spherical shape shrinks their body surface in comparison with their overall volume. Birds also have a cooling mechanism for their legs: the blood flowing into their feet loses warmth to the blood being pumped back up from their feet, lowering the temperature of these naked extremities to almost 32 degrees Fahrenheit, which is why waterfowl can paddle around in icy ponds without feeling pain.

Despite these adaptations, the smaller the animal, the larger the surface of its body relative to its volume. So, per pound, a bear has far less skin than a small bird; therefore, per pound, it also loses far less warmth. This means that very small birds—for example, the goldcrest, which weighs only two-tenths of an ounce—have formidable problems producing enough energy in the form of heat. Incidentally, the delicate song of the goldcrest is a good way to check your hearing. Its frequency is so high that many people over the age of fifty can't hear it at all. (I can barely hear it these days.) Unfortunately, the bird's voice doesn't help it keep warm, and it must replace the energy it is constantly losing through its skin and feathers, or the little singer will soon freeze to death—which basically means that it has to eat all the time.

While bears sleep in the comfort of their winter dens, Chickadees, Robin Redbreasts & Co. are constantly searching for calories. Unfortunately, there are often not enough to go round. Beetles and flies have retreated deep under the leaf litter on the forest floor or are dozing in the dead wood of downed trees. And berries and seedpods are either buried under snow or have already been eaten. No wonder many birds starve, most in their first year of life. The average life expectancy for European robins is little more than twelve months, even though the birds are capable of living for four years or more—if they have enough to eat.

When you see a small feather fluffball sitting there freezing in your garden, don't you feel sorry for it and feel you must do all you can to help? I was very strict for the first fifteen years we lived at the forest lodge in Hümmel. Feeding birds meant interfering, which meant changing their food situation in ways that were not natural. When you install a bird feeder and provide grain and fat, you promote the population of specific species of birds. Many of the young survive the winter, and the next spring these species are particularly populous—at the expense of others that perhaps did not come to the feeder. There's also the fact that in nature, reproduction rates are perfectly calibrated to winter losses. Species that lose more individuals over the winter simply lay more eggs and breed more than once a season.

So, is it okay to interfere? For years, I refused to, despite pleas from my children. In retrospect, I regret my decision. About ten years ago, I gave in and built a bird feeder. I put it out in front of the kitchen window so that we could begin our observations over breakfast. My wife, Miriam, and the children were delighted, and soon there was a telescope and a bird guide near the window, as well.

The most exciting moment came when a surprise guest showed up: a middle spotted woodpecker. I particularly love this bird because it is associated with ancient deciduous forests. Their habitat is threatened, because for these birds to feel at home, the beech forests need to have been around for a very long time. One reason for this sounds pretty basic: beeches that haven't reached the two-hundred-year mark have smooth bark. It's only with increasing age that they develop wrinkles and folds, as older people do—and only then can these woodpeckers find purchase on their trunks. This species of spotted woodpecker is also not very keen on constructing nesting cavities. Perhaps, unlike other woodpeckers, they get a headache from hacking away at wood.

Whatever the reason, this woodpecker either uses nesting cavities vacated by other species, or, if it absolutely must resort to its own handiwork (or beakwork, I should say), it works on trunks that are rotten, where the wood is already nice and soft. And this shy, rare bird was now at my bird feeder. Until then, I had assumed that there were no middle spotted woodpeckers in my forest, so I was doubly glad: once for the bird and once for the forest. The presence of this species is like an environmental seal of approval—and I'd been awarded one out of the blue. Of course, ever since then, I wait eagerly for these special ambassadors of the woods to appear—which they do regularly, because middle spotted woodpeckers are one of the few species of birds that remain tied to their territory even in winter.

Despite the joy of experiences like this, I want to re-revisit the question of whether feeding birds in winter is beneficial from an ecological perspective, because it certainly changes the rules in the bird world. One of Gregor Rolshausen's research teams at the University of Freiburg showed just how much. They were researching two different groups of blackcap warblers. The birds, which are about the size of a chickadee, are easy to identify. They have gray plumage and sport a cap on their heads: black for males and brown for females. The birds spend their summers in Germany, and in fall they fly to warmer places, such as Spain. There, they eat berries and fruit, including olives. In the 1960s, they established a second migration route that led north to the United Kingdom. The reason is that the British are great bird lovers, and they feed the birds in their country so well that they no longer want to fly south. The migration routes to this island nation are considerably shorter than the routes to Spain.

Bird food and olives are so different that the original shape of the birds' beaks is not the best for their new food source. Therefore, over the past few decades, the population of blackcaps that fly to the United Kingdom has begun to change—both visually and genetically. Their beaks have become narrower and longer, while their wings have become rounder and shorter. Both developments are adaptations to life at the bird feeder, because the redesigned beaks make it easier to pick up seeds and fat. The wings are no longer ideally suited for long-distance flights, but they improve the maneuverability the birds need for short flights in the garden. And because this population and the population that flies south rarely interbreed, a new species is gradually being created—a new species that is forming because of winter feeding.

You could call this a massive interference in nature, but is it necessarily negative? You could consider the development of a new species something to be celebrated. After all, increasing the overall number of species is always a win for an ecosystem, and in this case the new species means the birds have adapted to better fit a changing environment. It becomes a problem, however, when an altered species then interbreeds with the original species, changing the genotype so much that the original form of the blackcap, for example, might cease to exist.[2]

We can observe changes like these in many culturally manipulated plants, including fruit trees. There are hardly any genetically pure wild apple or wild pear trees left; they might well be on their way to dying out completely. The cause is the relationship between people and fruit, which stretches back for thousands of years and is intertwined with the history of fruit breeding. Because bees don't care which blossoms they pollinate, they carry the pollen of fruit trees bred by people to the blossoms of their wild counterparts. The genetic material of the two gets mixed, and the offspring of the wild trees are changed in the process. At some point, pollinating insects will wipe out the last of the wild fruit trees, and all the trees will be hybrids. Is that important? We don't know, but we do know that something irretrievable is being lost. This happens in the animal world, as well. An auroch looks out at us from behind the eyes of every cow, albeit—genetically speaking—from a great distance. Although it's impossible to revive aurochs in their pure form, Heck cattle—which have been bred back to have at least a visual resemblance to aurochs—now graze in a few nature preserves in Germany.

BUT, OF COURSE, there are completely different reasons to feed birds, and here I come back to those emotions I mentioned earlier. It's not just the middle spotted woodpecker that has shown me how much enjoyment you can get from birds but also Koko the crow. I introduced Koko in my book *The Inner Life of Animals*. We see the crow only in winter, and when we do, it's all about food. Our two horses, Zipy and Bridgi, are out in the field year round, because fresh air helps keep them healthy. They're old ladies now, and they get a handful of grain every day so that they don't get too skinny. Koko used to peck undigested seeds out of their droppings, which I found less than appetizing.

For some years now, my wife and I have been putting out a few kernels of grain on the rail where we tie up our horses so that Koko can get a somewhat more hygienic breakfast. I completely overlooked the fact that the crow was communicating with us nonverbally. One day, it flew past me with an acorn in its beak and hid it in front of me in the grass. When it realized I was watching, it retrieved the acorn and flew on a bit farther so that it could bury the nut safely out of my sight. Only then did the crow fly back and collect its morning portion of grain. When I recounted the story at the breakfast table, my children enthusiastically suggested that I include it in my animal book.

You'd think that I'd be better prepared to notice the bird's behavior after that, but unfortunately, that was not the case. And so it was that I completely missed it when Koko left us a token of his affection. I only realized what the crow was doing when Jane Billinghurst got in touch with me. She had already translated *The Hidden Life of Trees* into English, and she was working on *The Inner Life of Animals*. To make my account more immediate for English-speaking readers, we were replacing some of the German references with similar ones published in English. Among her other suggestions, on the subject of gratitude (and whether and how animals express this), Jane suggested mentioning a BBC report about a story out of Seattle.

The story was about a girl name Gabi. When Gabi was four years old, she would occasionally drop some of her food on the ground by accident when she was eating outside. In no time, crows snatched up the unexpected gifts. Later, Gabi got into the habit of sharing the contents of her school lunch box with the crows, because she liked them. Finally, she began putting food out just for them. She set out containers of nuts, provided water, and scattered dog food in the grass. This was the turning point in the relationship between Gabi and the crows, because now the crows began to bring Gabi presents, as well. They brought her little pieces of glass, bits of bone, and small beads or screws. They left their gifts in the containers after they had eaten the food. Over time, Gabi amassed quite an astounding collection.[3]

I found the story moving and quickly agreed to mention it as an example of animal gratitude from North America. The next time my wife and I trudged out to feed the horses (it was in December), we noticed a small apple on the rail. And that was the first time I realized what was happening. Koko had been leaving us gifts for years; we just hadn't realized that this was what the crow was doing. We had often puzzled over the fruit, stones, or sometimes bits of mouse that were laid out in the same spot where we had left the food, but it had never crossed our minds that these were gifts intended for us. In hindsight, we're sorry that we hadn't noticed earlier what Koko was trying to communicate, and these days, we're especially excited when the crow leaves us something.

And so, yet again: is feeding birds at feeders bad for them? Aren't we interfering with natural processes when we do this? Without our help, Koko might have starved long ago, and another crow or another species of bird might have filled the vacancy Koko left in the ecosystem. We've already covered the pros and cons of the direct impact on the environment, but we haven't talked about another aspect here: empathy. Empathy is one of the strongest forces in conservation and can achieve more than any number of rules and regulations. Think of the campaigns against whaling or against the slaughter of seal pups—public outcry was so loud only because we all empathized with the animals. And the closer the animals are to us, the greater our empathy.

And I mean that literally, which is one reason I'm not opposed to zoos in principle, as long as the animals are looked after in a species-appropriate manner. People who experience animals up close feel more strongly connected to them and are more prepared to do something to protect them. That's why I think it's a shame that private individuals (in Germany, at least) are not allowed to own wild animals. For species that are not threatened by extinction, on balance, this could do more good than harm. Anyone who has had the kind of experiences I have just described will no longer curse the magpies in the front garden or support the shooting of ravens. Certainly an animal will occasionally be loved to death because it's not being looked after in the right way, but in the end, the best way to protect nature is to ensure that people experience it.

One small tip, by the way. Birds don't need just food in winter; they also need water. A little ice-free water in a dish can sometimes be even more helpful than bird food. We see this at our horses' drinking trough. The horses live outside in their pasture all year, even on freezing cold days. They much prefer this, I hasten to add, to standing in a warm stable. Anyway, water is a problem because the trough keeps freezing over. The only solution is buckets of warm water that we carry out either in a wheelbarrow or on our ATV. And from time to time we see Koko and his companions helping themselves to a sip of cool, clean water from the trough after their meal of grain.

With other animals, however, feeding in winter is not a good idea, because it can have a negative impact on the whole ecosystem. How that happens and why trees today will never be free of wild boar is too much to fit into this chapter. So, let's begin a new one.

10

How Earthworms Control Wild Boar

I'VE OFTEN HEARD it said that mild winters cause plagues of biting flies or catastrophic bark beetle infestations. I've already explained that explosions in bark beetle populations are more often caused by commercial forest practices, but I think it's worth taking a closer look at the impact of winter in the forest. Severe winters are characterized by weeks of hard frosts and at least some snow cover. Everything freezes. The top couple of inches of the soil are rock solid, and life in the cold seems to be anything but a walk in the park.

Let's start by taking a look at how winter conditions affect smaller animals. Insects exploit laws of nature to protect themselves against freezing. They use sugars they produce naturally to create a kind of antifreeze, and they empty their gut to minimize their water content, because tiny amounts of water don't freeze until temperatures fall far below 32 degrees Fahrenheit. Five microliters of water (which is a vanishingly small two ten-thousandths of a fluid ounce), for instance, doesn't form ice crystals until it reaches 0 degrees Fahrenheit. Despite this, the youngest bark beetles struggle to survive. If it stays cold for too long, larvae pass on to insect Nirvana without seeing spring. Their lives end not because they can't survive freezing conditions, but because water has entered their mouths and breathing tubes. Although fluids inside the larvae are protected against freezing temperatures, water entering from outside their bodies freezes immediately when the temperature falls. That's why the little ones survive particularly well when a thick layer of snow keeps out the worst of the cold. Because adult beetles don't have this problem (they are good to −22 degrees Fahrenheit), bark beetles try to avoid breeding in fall.

Mild winters are catastrophic for bark beetle larvae, as well, because mild means damp. Think about it. What kind of weather would you prefer to be out in? A few degrees above freezing in rain or a few degrees below freezing in the sunshine? I, for one, would prefer the latter. If temperatures are below freezing, you usually stay dry, which means it's easier to keep warm. Above 40 degrees Fahrenheit, moisture-loving fungi become active again. They get to work inside the overwintering insects and start eating them while they dream.

WHEREAS OVERWINTERING BARK beetles hold still and wait for spring, most mammals are awake and active in winter, which means they also need to eat all the time to maintain their body temperature. This puts them in the same predicament as the birds. Shouldn't we take pity on our four-legged fellow creatures, as well? Shouldn't we feed them? We're already doing that—at least with some species. Have you ever seen a fodder rack in the forest? Or perhaps a couple of wooden boxes filled with feed corn? This food is supposed to help starving roe deer, red deer, and wild boar survive the winter. We also know that this supplemental feeding is not an act of altruism. This provisioning is done only for animals whose antlers or tusks would look good displayed as hunting trophies over the living room sofa. No thought is given to animals such as foxes or squirrels. But that wouldn't be necessary, anyway, as they are well adapted to their climate and have developed their own strategies to survive the cold months of winter.

Whereas squirrels cache food and sleep a lot in winter, deer have adopted a different strategy: they regulate their body temperature while spending the cold months of the year standing around dozing in the undergrowth. Researchers at the University of Vienna have discovered that deer can lower their subcutaneous temperature to 60 degrees Fahrenheit to save energy—which is amazing for large warm-blooded animals. According to project lead Walter Arnold, this strategy is similar to hibernation.[1] Using this method of saving energy, deer can make the fat reserves they put on in the fall last into the following spring. It's just the weak or sick animals that starve, which is a natural method of ensuring the genetic health of the species.

Winter feeding can lead indirectly to starvation, particularly with red deer. This happened in Germany in the winter of

2012–13, when a lot of snow fell everywhere in the country. In the district of Ahrweiler, where I live, the deer population had increased so much that the animals were practically tripping over each other in the forest. Hungry deer were appearing in cattle shelters on farms and eating the cows' feed. A colleague of mine even sent me a photograph of a doe having a snack at a bird feeder. Of course, the hunting community began calling for a permit to feed them. Hunters even visited schools to solicit sympathy for the animals and gather support to influence politicians.

Many dead deer were found, which heated up the debate: Were we really going to allow these noble animals to starve? When veterinarians examined the animals, however, they made an unexpected discovery. They found the victims' stomachs were stuffed full of food, which ruled out hunger as the cause of death. They also found enormous numbers of gut and stomach parasites, and these were what had sealed the deer's fate.[2] Because of the increased numbers of deer, the animals had more contact with each other and with contaminated feces, which helped parasites spread widely. Once ingested, the parasites, not the deer, were the beneficiaries of the nutrients in the deer's guts, which meant that many deer weakened and died, even though they had plenty to eat—death by starvation was an indirect consequence of the feeding program.

The findings didn't change the hunters' views, however. As far as they were concerned, it was still better to have as many of the large plant eaters as possible survive, because that meant there would be something to see every night when they perched in their hunting hides. Overpopulation, however, also leads to stress as deer fight over territory, and stress in wild animals leads to reduced body weight and, especially in roe deer, smaller antlers. That is an unintended consequence, because what hunters want is lots of wild animals with huge trophy racks. Ignoring what is really going on, hunters in Germany continue to try to fatten up weak individuals, which makes things even worse. And feeding deer isn't cheap.

The journal *Ökojagd* once estimated from hunters' reports how much they fed game. They came up with 12.5 pounds of corn per pound of bagged game.[3] That is way more food than the meat industry uses in the mass production of meat. And because nature is the way it is, food is immediately converted into reproduction, so the number of individual animals explodes. The result is wild boar in vineyards, in front gardens, or even on the Alexanderplatz, a large public square in the middle of Berlin, because the forest is gradually becoming too cramped for the number of game animals out there.

HUNTERS' INTERVENTIONS IN the finely calibrated balance of nature create more losers: trees. Over millions of years, trees have developed a perfect strategy to keep large browsers away, but this strategy no longer works when the animals are being fed.

The two most important native trees in Germany, the beech and the oak, produce very large seeds. A beechnut weighs only two one-hundredths of an ounce, but that's quite a lot for the seed of a forest tree. Spruce seeds—which are an important food for squirrels, mice, and many birds—are four one-hundredths the weight of a beechnut, but they are still very attractive to animals. Beechnuts, however, are what you might describe as real calorie bombs, partly because of their size, but also because they are almost 50 percent fat. Acorns are even heavier, with an average weight of fourteen one-hundredths of an ounce. The fat content of an acorn may be only 3 percent, but the carbohydrate content is 50 percent, making acorns the top prize in the fall food lottery.[4]

This lottery plays out on a three- to five-year cycle. In the off years, hunger is the order of the day for many animals, and that is exactly why beeches and oaks don't produce fruit every year. The breaks they take ensure that populations of wild boar, roe deer, red deer, birds, and hordes of hungry insects don't come to depend on their bounty.

Wild boar, in particular, are really good at snuffling up these desirable seeds, and some years they eat every last one of them from the forest floor. Boar populations can quickly triple, and a year later, large groups of young pigs roam through the fall leaves, overturning every twig, stone, and rotting stump. The following spring, no new beech trees sprout and no young oaks see the light of day, and if this situation persists for decades, forests begin to age.

When an ancient tree dies, grasses and bushes grow in the clearing that opens up around it, gradually creating a small grassy plain where browsers move in to eat all the small trees that try to sprout. Trees, however, know how to put a stop to that, and one way they put the brakes on meadow making is to leave long intervals between the times they bloom in order to reduce the number of browsers in the neighborhood.

But that's not all. What use is it if only some of the trees take a break while others are loaded with beechnuts and acorns? Wild boar go hungry only when there are no nutritious seeds anywhere for at least a few years.

And so the trees come up with a communal bloom strategy, and all trees of the same species have to be on board. It's not enough, for example, for one stand of beeches to come to an agreement via their root and fungal connections in the ground. That form of communication works well (and, astonishingly enough, some of it is via electrical impulses), but the wood wide web doesn't reach far enough for this particular purpose, because wild boar can roam long distances and simply find another stand of trees 6 to 12 miles away. Therefore, trees agree among themselves over long distances, and by long distances I mean hundreds of miles. We don't yet know exactly how this works, but the important thing is that, with the exception of a few rogues, all the trees in large areas synchronize when they form fruit and when they take a break from reproduction.

In Germany right now, the deciduous trees' strategy is being completely obliterated by hunters. They feed wild boar not only in winter but often year round, which cancels out the food shortage planned by beeches and oaks. As part of their work, foresters in Baden-Württemberg examined the stomach contents of wild boar that had been shot. This led to the discovery that on average through the year, 37 percent of the food they eat comes from hunters. In winter, this percentage increases to 41 percent, and this is especially dangerous for the wild boar,[5] because in the cold months of all years except mast years the forest is mostly empty of food—as you would expect the stomachs of the wild boar to be. Without the hunters' intervention, many of them would starve, and then the population would match the carrying capacity of the habitat once again. But this is not what happens when they never have to adjust to periods of scarcity. The boar can help themselves anytime at one of the thousands of feeding stations, which further boosts their rate of reproduction. The German Ecological Hunting Association (ÖJV) once calculated what that meant in concrete terms for individual animals: in extreme cases in the Westerwald mountains, it came to as much as 1,700 pounds of food per shot sow.[6]

The hunters try to conceal the true causes of higher populations. They blame farmers and their huge hog-heaven fields of corn. They argue that climate change, with its warmer winters, favors dramatic increases in population. They point out that hunters no longer feed wild game, because the practice is mostly no longer allowed, at least for wild boar. That's true, because the word "feed" has been replaced with the word "bait." We're talking about a small amount of feed corn put out in a clearing to lure animals within range of hunting blinds, where they are shot. Thus, according to the official version, baiting leads to the reduction of wild populations and not to their increase. However, there is so much baiting going on that the population of wild boar is increasing faster than it can be shot; consequently, baiting is not having the stated effect, and the whole situation is getting out of hand. To say nothing of the fact that in most districts, illegal feeding continues on a massive scale.

Out there in the forest, mostly out of public view, just about everything has been dumped that wild game might find tasty. At the beginning of my career, I found whole truckloads of tulip bulbs in a clearing. They were clearly not saleable and therefore needed to be disposed of. Local hunters probably thought, "Why not make a virtue out of necessity?" and had the cargo transported to the forest. The wild boar seemed to have enjoyed it, because after a few weeks the bulbs had disappeared.

Apples that are too small or too light according to European Union regulations or that don't look the way apples should are also disposed of as food for wild game. An acquaintance told me that people with hunting leases in her village once scattered tons of pralines that looked mouth-wateringly fresh. Hunters are basically behaving the way owners of large restaurants used to behave decades ago, when it was normal to keep a barn full of pigs to process leftovers and recycle rejected chicken stew, duchess potatoes, or pork and beans into fresh meat. It's the same when you feed wild game in the forest. The only difference is the pigsty, which is much larger and made of trees.

Meanwhile, the relationships that used to exist in ancient forests have been completely upset by foresters and hunters. Whereas there used to be very few roe deer per square mile, today there are, on average, 130. As red deer are animals of the plains, they were hardly ever encountered in the forest. Now, in many forests, there are about twenty-five red deer and twenty-five wild boar per square mile, too, which means it's really crowded. It's become a real zoo out there in the forests of Central Europe, which makes the heart of every hunter beat a little faster.

If there are hordes of browsers demolishing most of the next generation of trees, do our deciduous trees have any kind of a future at all? We mustn't give in to such gloomy thoughts. Luckily, it's only a matter of time until conditions improve. For one thing, there are the wolves, which are slowly returning all over Europe to set things back on track like they did in Yellowstone. For another, the trees have more secret allies. Surprisingly, one of these is a creature that lives in the soil: earthworms can be very dangerous for wild boar. Earthworms? Don't they stay peaceably in their tunnels, munching fallen leaves and excreting them as humus? That's right, but earthworms still pose a danger to pigs. (This applies in Central Europe, anyway, where earthworms are a native species. Earthworms were wiped out in many forests in northern North America in the last ice age, and different species of decomposers have evolved there to recycle forest waste.)

But, back to Europe, where wild boar plow through soft ground with their plate-shaped snouts searching for meat, and one of the best food sources is earthworms. Per square mile, up to 850 tons of them can live beneath the soil surface.[7] To put this into perspective, the weight of all the large mammals living in a similarly sized area (roe deer, red deer, wild boar) comes to only one-third of that. Incidentally, in case of emergency, it would therefore also be much more effective for people to plow the land in search of earthworms instead of going out hunting.

Back to the pigs. They eat earthworms, which in themselves are harmless, but the wild boar help themselves to stowaways as they do so. These stowaways are lungworm larvae, which develop inside the earthworms while waiting for a suitable host for the final phase of their life cycle. That host—and here we touch briefly again on that emergency situation I mentioned—could be a person. This means if you do end up having to eat earthworms, you should cook them thoroughly first. If a wild boar eats a meal of lungworm larvae, the larvae travel through its bloodstream and end up in its lungs. There, the larvae settle in the boar's bronchial tubes, where they mature into adults that cause swelling and bleeding. The wild boar then excrete the eggs, earthworms ingest them, and the cycle is complete.

Thanks to their weakened respiratory organs, the bristly porkers become susceptible to a whole armada of other illnesses, and there is a higher risk of mortality, especially among piglets. The higher the numbers of wild boar, the more earthworms carry lungworm larvae, which, in turn, means more infected pigs. The whole thing spirals upward until, at some point, the pig population crashes. Fewer animals = fewer excreted eggs = almost no infected worms. Lungworms, therefore, regulate wild boar populations, but there are still more little adversaries out there that the pigs have to deal with.

A whole army of infectious agents has its eye on wild boar, including a large number of viruses. Viruses are remarkable, but what exactly are they? Scientists don't include them among the living species of this earth, because they have no cells and can't reproduce or metabolize on their own. All they are is a hollow shell that contains a blueprint for multiplication. Basically, they're dead. At least, they're dead as long as they aren't docked onto an animal or a plant. Once they've made this connection, they smuggle their blueprint into the host organism, forcing it to create millions of copies of the virus. In the process, mistakes are always made, because viruses, unlike cells, have no built-in mechanisms to fix mistakes.

All kinds of mistakes, however, mean all kinds of new variations on the virus. It doesn't matter that many of them go nowhere, because there's always something useful in a junk pile. That's how viruses adapt quickly to new conditions and can attack their hosts more effectively. New mutations, especially, have the potential to kill. Normally, it doesn't make sense for a virus to kill its host, because after it attacked its host it wouldn't have an opportunity to proliferate in the future. Only fresh mutations make mistakes like that, because they have not yet adapted sufficiently to be able to exploit their hosts without killing them.

Naturally, the opposite is true for the host: a host that has a long relationship with a virus adapts over time so that the illness caused by the virus becomes relatively harmless instead of fatal. Chicken pox is an unfortunate example here. Europeans are well adapted to this illness, which usually strikes in childhood, but the virus—brought by white colonizers—devastated Indigenous populations in North America. Together with measles and other diseases, chicken pox killed up to 90 percent of the people in some areas.

Animals face the same predicament. Our global economy creates similar conditions for them that humans created when they arrived on continents that were new to them. Illnesses that native fauna don't know how to combat arrive in new territory packaged up with trade goods and in live animals and plants. African swine fever is one such disease. It's normally found in Africa, where bloodsucking soft ticks transfer the disease from one animal to another. The first place it was identified outside Africa was Russia in 2007, and it began spreading in Europe because people let it in. We don't know exactly who imported it, but it probably arrived in a shipment of pork that contained the virus. From there, the virus probably spread through the illegal disposal of slaughterhouse waste and carcasses. The death rate for infected animals is extremely high: in fact, it's 100 percent.[8]

Is this a dramatic development for wild boar? It definitely is for individual animals or for a pig family. Wild boar are social and love to cuddle up together, so the infection can jump from one animal to the next. Even if all members of the family are not infected, the whole family suffers. Wild boar love their parents, children, brothers, and sisters, and miss them when they die. However, for the evolution of the forest ecosystem, swine fever is not necessarily catastrophic. It's difficult for the disease to spread naturally in Central Europe, because there are no soft ticks here to act as intermediate hosts. Here, it is the unnaturally dense population of pigs that makes direct transmission from one animal to another so much easier. If the disease thins the population, there will be less contact between pigs, and the virus will not be able to travel farther. Then the epidemic will end, there will be fewer pigs trotting around in the forest, and beeches and oaks will have room to breathe again.

There are connections that we know to be true because they have been well researched (as in the case of wild boar and swine fever), and then there are connections that we assume to be true because they have been handed down for generations. But maybe the time has come to take a closer look at things many of us take for granted.

11

Fairy Tales, Myths, and Species Diversity

WE'VE LOOKED AT some surprising natural connections. Yet I haven't mentioned others that might seem much more obvious—for example, how events in fall predict the harshness of winter. And there's a good reason: they don't exist.

For example, since ancient times, the fruit set of beeches and oaks has been used to forecast the weather. An old German country saying goes something like this: "Many beechnuts and acorns and winter will be harsh." Or "Many acorns in September, much snow in December." To try to get to the truth behind these sayings, we first have to ask a few questions. Why would a tree do this? How could forming lots of seeds help a tree survive a harsh winter? What would the indirect consequences be for the tree?

Unfortunately, I don't have the answers to these questions. All we know is that the trees in each species (oak or beech) agree on a time to bloom together so that every few years they produce an enormous amount of seed. As I mentioned earlier, they do this to regulate the size of the plant-eating population by ensuring that browsers can't rely on a constant supply of food every year. But the strategy has nothing to do with winter.

There's another point to be made here. Flower buds (just like leaf buds) are set the previous summer. If a tree matched its seed production to the winter temperature, it would have to know more than a year in advance what this was going to be and plan accordingly. Beeches and oaks, however, have no better ways to predict winter weather than we do. What trees can do is register shorter days and falling temperatures. They use this information to decide when to drop their leaves before the first heavy snowfall. And many years they don't even get this short-term forecast right, as you can see when winter arrives early. When snow falls in October, as it often does, branches with green leaves still on them break under the heavy weight of fresh wet snow, which is a painful lesson for the trees. At least if this happens to them when they are young, they can learn from their experience and drop their leaves a little sooner in the future. But they only drop their leaves early as a precaution: their decision doesn't have anything to do with improved forecasting. So there you have it: even beeches can't forecast the weather a year in advance.

So much for representatives from the plant realm, but what about animals? There's also folk wisdom that says squirrels can predict a harsh winter. If they are particularly busy gathering food and laying up large stores of acorns and beechnuts, that means the winter is going to be particularly harsh. Really? I think you could probably answer this one for yourself. Of course these pretty little creatures don't have a sixth sense for what the weather will be like in the coming months any more than trees do. Their drive to collect nuts is simply a question of supply. When trees produce a lot of nuts, the little rascals hide a lot of them. In the years when the trees have agreed to take a break and there are barely any nuts on them, the animals find far fewer of them and we're unlikely to see them squirreling them away.

Then there are those folk sayings that straddle myth and reality. They express relationships that exist, but the explanations they give are incorrect. The classic one for me is the combination of ticks and broom. The conventional wisdom holds that these little bloodsuckers like to live in broom. This shrub is widespread everywhere in Europe where the Atlantic ensures cool summers and mild winters, as it does in the Eifel, where I live. Broom is so common here it defines parts of the landscape. In spring, the shrubs are totally covered in golden yellow flowers that look like butterfly wings. The blooms are so thick, there's not a spot of green to be seen. Large broom hedges infuse the landscape with a golden glow, and that's why the plant is called Eifel Gold.

So, do ticks really like broom? All parts of the plant are poisonous, and not only for people. Substances in the branches, flowers, and leaves discourage browsing, and roe deer, red deer, and cattle generally give the hedges a wide berth. Where there's a dense population of wild game, most of the other, tasty shrubs get eaten. Broom, therefore, has an advantage over the competition and can spread unchecked—and that's just what it does, both stubbornly and successfully. This shrub has developed a number of different strategies for seed transport alone. In the heat of the midday sun, for example, the seedpods burst open with a loud crack, scattering seeds in all directions. The round seeds roll down slopes easily to travel even farther.

Broom doesn't stop there, however. It also employs ants. Yes, those secret sovereigns, yet again. Ants distribute the seeds and help Eifel Gold get established in every corner of the countryside, even in forests. It's too dark there for the broom seeds, but time is on their side. They can lie in the humus for more than fifty years, until one day a storm or a human entrepreneur fells the trees. Rays of sunlight hit the forest floor and gently wake the slumbering seeds. They quickly sprout and grow to be bushes that are nearly 2 feet tall in their first year. Only young trees or other bushes, such as raspberries, cramp their style, but munching roe deer come to their rescue. The deer quickly eat the fresh green growth, removing the annoying shadows falling on the young broom bushes.

While the deer are munching, a special sort of hitchhiker tumbles from their fur: ticks. The ticks that fall are particularly large ones in the final stage of their life cycle. They have been sucking up blood one last time before loosening their grip and falling off so that they can crawl into the nearest bush to lay their eggs and die. The young ticks hatch, are scraped off their bush by passing mice, and pick up where their mothers left off: sucking blood. Then they, too, let go and fall off after their meal, grow, and molt. Finally, they hang out, hungry once more, in the surrounding vegetation—in broom, for example—and wait there for larger mammals (and, perhaps, for people). And that is why there are always large numbers of ticks where there are large numbers of deer, and it is deer that ensure broom can spread freely. What ticks love is not broom but warm-blooded hosts. Broom is simply a species that also profits from the presence of browsers, and broom and ticks appear together in habitats where there are overly high numbers of deer. The ticks and the broom depend on the deer but not on each other.

TREES CAN ACHIEVE great things together without meaning to, even when their achievements have nothing to do with survival. Every year in fall, a drama plays out that makes me think of a children's playground—especially the merry-go-round. Do you know what I'm talking about? Someone spins this flat disk with bars on it, while the children sitting on it stretch their legs out in front of them away from the center. When the children draw their legs in, the merry-go-round spins noticeably faster; when they stretch them out, it slows down again. It's doubtful trees have fun on merry-go-rounds, but they do something similar every year. When deciduous trees in the Northern Hemisphere drop their leaves all at the same time, we all spin a little faster and the days get shorter. You don't believe me?

We're talking only tiny fractions of a second, which, I have to say, are barely noticeable because of the influence of other global processes, but the change is measurable. Most of the landmass of the earth lies in the Northern Hemisphere. Therefore, most of the trees are here, too. When deciduous trees lose their leaves, their foliage ends up 100 feet closer to the center of the earth (the difference between the treetops and the ground). The effect of this shift in weight toward the center is similar to the effect of children drawing in their legs. In spring, when new leaves grow, exactly the opposite happens. The fresh, water-filled leaves shift a lot of weight up higher or, to put it another way, farther away from the center of the earth, slightly slowing us down again. A more fun way to look at it is that thanks to the trees, we get to ride on a merry-go-round. However, because the effect makes a difference of only fractions of a second, and because there are other processes that affect the earth's center of gravity, such as tidal currents, we may as well place this spinning phenomenon in the basket of half-truths that blur the line between fact and fancy.

THERE'S A VERY different kind of myth surrounding species diversity. When we save individual animals or plants, we really believe we're doing something good for the environment. Yet this is rarely what happens, mostly because when we have to change conditions in the environment to ensure the survival of one species, the survival of many others ends up in jeopardy. But I'm getting ahead of myself.

When we see just how multilayered the interactions between different species are, we have to ask, once again, whether we will ever be able to fully comprehend the connections in our environment. The examples we've discussed so far involve just a few animals influencing each other in highly complex ways. Imagine a juggler who's got two balls in the air. Every time a new species enters the act, things become much more complicated and difficult to follow. According to current estimates, there are 71,500 of these "balls" in Germany (including animals, plants, and fungi), and 1.8 million species have been described globally to date.[1]

That sounds complicated, and it's even more complicated than it sounds, because there are many animals and plants we haven't discovered yet. The World Wide Fund for Nature and the Mamirauá Institute for Sustainable Development have reported that in 2014–15, a new species was discovered in the Amazon every 1.9 days. Recently, I was talking with a researcher who, incidentally, is herself a member of an endangered species: entomologists. There's simply not enough money to research beetles, flies, and other insects, she told me, and, more importantly, not enough young scientists entering the field, which means that even in Germany there are still a good many blank spots on the species map. To the 71,500 known species in Germany, therefore, you can add an unknown number of species whose effect on the ecosystem is, of course, unknown. Clearly, we can't even understand everything about nature right here where I live—but, in my opinion, we don't need to.

In the examples laid out in previous chapters, it's easy to see how fragile the system is and what the consequences are when just a single species falls off the map. Knowing this, we must strive as much as we can to preserve intact landscapes or allow them to manage on their own. But what does intact mean? Whom should we trust here?

In Germany, forest agencies and owners claim that commercial forests are good for species diversity. The third National Forest Inventory, published in 2014, revealed that the average tree is now seventy-seven years old—bravo! The brochure, published by the Federal Ministry of Food and Agriculture, also extols the ecological importance of old trees and indicates that everything is fine on this score.[2] The tree sap hoverfly, *Brachyopa silviae*, would certainly dispute this if it could. This tiny fly was first discovered in 2005. It has only been spotted six times in the world, so you could say it is extremely rare. And there is a reason for this.

Even though it has wings, it seems that this fly doesn't travel very far and prefers to remain in undisturbed forests, where it feels at home. Here, the fly finds wounds under tree bark that are weeping sap—its favorite food. Or, I should say, this weeping sap provides the substrate for its favorite food, because the sap is food for bacteria and other microorganisms that form a slimy mat on top of it, and this slimy mat is where the tree sap hoverfly likes to graze. But weeping patches like these can be found only in trees that have at least 120 years under their belt (or, perhaps I should say, their bark). At least 120. The trees can be older, of course, but if federal promotional literature is happy with an average age of 77, you have to get alarmed and worry about this fly.

Dr. Frank Dziock discovered this fly only by chance.[3] He had set out insect traps in flooded areas to catch hoverflies, because he wanted to find out how they reacted to high water levels. At first, Dziock didn't realize that he had found something remarkable in his trap. Then he noticed two spots on the back of one of the flies. No other known fly sports these spots, and so what he had here was an as-yet-undiscovered species.

This fly needs wounded old trees. Old trees, however, are highly vulnerable in commercial forests because of thinning, which targets damaged trees. The long-term goal is to let only perfect specimens of beech and oak grow larger and older so that their valuable wood can be harvested. Too bad for the flies; their needs are completely ignored. It's true that a few trees are left scattered about for environmental reasons, but if all the others around them are chopped down, these holdouts won't grow to be very old. They will be struggling along without the typical damp, cool microclimate of the forest, and direct sunlight will be heating up the ground all around them. In addition, their root and fungal networks—which help support old and sick trees—will have been destroyed. This network is crucial for forest health, so let's take a closer look.

IN MY BOOK *The Hidden Life of Trees*, I talked about the forest internet—or the wood wide web, as the journal *Nature* fittingly described it. The network is made of fungi. They grow their filaments through the soil and connect trees and other plants with each other. Fungi are remarkable fellows. They don't belong in either the plant or the animal kingdom, but they have much in common with animals. Photosynthesis: nope. They have to get their food from other living entities. Like insects, they have chitin in their cell walls. A few of them—slime molds, for example—can even travel from one place to another. Not all of them are friendly, however. Honey fungus, for example, attacks trees to get at the sugar supplies and other delicacies stored inside. It often kills its victim and then moves along the ground to the tree's next family member. Related trees are not defenseless in the face of attack by fungi or insects; they heed warnings from other trees, including scent signals that contain information about which villain the trees are up against. The appropriate defensive compound can then be stored under the bark to spoil the appetite of hungry insects or mammals.

Unfortunately, the wind often blows airborne warnings in only one direction. This is where the roots come in. They connect with the roots of other related trees and transmit important news bulletins via both chemical and electrical signals. But this root network can't reach every corner of the forest, and sometimes the connection is broken when an ancient tree dies.

Fungi help bridge such gaps. Like the fiber-optic cables of our internet, their subterranean filaments carry messages from tree to tree so that the whole forest soon knows what to expect. However, this service isn't free; the fungi tap Beeches, Oaks & Co. for up to one-third of the sugar and other carbohydrates they produce by photosynthesizing. That is a sizeable chunk of energy and is about the same amount the tree uses to grow wood. (The other third is converted into bark, leaves, and fruit.)

Anything that makes such high demands must also deliver dependably. And fungi seem to have mastered this, even though it's a tricky task. The wood wide web often suffers massive disruptions. For example, in winter wild boar roam the forest plowing deep furrows in the ground as they search for beechnuts, acorns, or the nests of mice. Inevitably, this activity disrupts fungal connections for many square yards. No problem for the fungi. As insurance, they extend many filaments parallel to each other, and they simply switch the connection to neighboring threads. Incidentally, that's why when you're out collecting ceps, boletes, or chanterelles in the fall it doesn't matter whether you twist the mushrooms or cut them off (a perennial bone of contention among nature lovers). Any damage is quickly bypassed underground.

In addition to sharing information and transporting sugar from one tree to another to support weak members of the tree community, the fungi also help trees reach essential minerals. For example, when trees suck up phosphorus compounds, they soon exhaust the supply within a few fractions of an inch of their roots. So it's a good thing that fungi connected to the larger network spin their filaments around the trees' delicate feeder roots to extend their reach. That way, all the nutrients the trees need can be delivered to their door even from distant parts of the forest floor.

Fungi can live to a ripe old age, but like every living thing, they start very small: as spores. And spores have a big problem. If they drop directly down from the cap of the fruiting body (the mushroom), they fall onto ground already occupied by their mother, which means they're not colonizing new territory. Billions of microscopic specks fall out of a single cap with travel on their minds—which is a real problem on a forest floor where normally no breezes blow. And this is where the special construction of the fungal fruiting bodies comes into play.

The fruiting bodies are mostly constructed as a stalk with a cap on top, and there's a good reason for that, as biomathematician Marcus Roper at the University of California, Los Angeles discovered. The spores trickle out of openings on the underside of the cap, where they are sheltered from the rain so that they don't get wet and clump together. The cap itself transpires water vapor, cooling the air around the mushroom just a little bit. The cooled air around the edge of the cap sinks slightly, taking the spores with it, and then warms up again in the surrounding air. Both the warmed air and the spores now waft out and up about 4 inches above the cap.[4] All it takes is a tiny breeze to carry the tiny passengers away, and the survival of Ceps & Co. is assured.

With any luck, one of the tiny spores lands on ground that is not yet spoken for. There, it extends a few of its small threadlike filaments (hyphae) and waits for signals from plant roots. If there's no chemical call in the neighborhood, the spore retracts its hyphae. It has enough food for multiple attempts at connection.[5] If it succeeds in making contact with the plant it's looking for—a beech tree, for example—a long life can begin, a very long life. Fungi can be every bit as long lived as trees. Ancient honey fungus networks have been found underground in North America. The record holder is a fungus belonging to the species *Armillaria ostoyae*. It is 2,400 years old and has spread to cover 3.5 square miles.[6]

We've only begun to scratch the surface when it comes to researching the world of fungi, and there are untold numbers of secrets hiding under every step you take outside. But there are also industrious creatures in the trees that require very specific conditions to survive. No, I'm not talking about bark beetles, which are simple souls when it comes to their dietary needs. They usually require just one thing of trees: they must show signs of weakness so that they can't defend themselves. If that's the case, all the beetles need to do is start munching on their bark and cambium—the layer of living cells between the bark and the wood. And because these conditions are almost always met in the range of any given tree (and the bark beetle that specializes in eating it), there are hardly any endangered species of bark beetle. Things are quite different, however, with specialists. They are so demanding, you could almost call them picky. And as demonstrated by the case of *Tenebrio opacus*, a kind of mealworm, "picky" might be an understatement. This insect feels comfortable only after a whole list of requirements has been checked off.

Let us say that once there was an ancient beech forest where a pair of black woodpeckers had taken up residence. They claimed many square miles of forest as theirs and fashioned an extended series of living quarters. Because the wood was hard and difficult to work, the birds took their time. Unlike other woodpeckers, black woodpeckers prefer to set up house in healthy trees—would you want to live in a house that was crumbling about your ears? Healthy beeches, however, are very hard, even for woodpeckers. In contrast to a human brain, a woodpecker's brain sits firmly in its skull so that it doesn't bounce back and forth while it's using its beak to deal staccato blows to a tree. As an added precaution, there's a special springy support behind its beak that cushions the blows before they travel to its skull. Despite this, fresh wood is simply too dense. But the black woodpeckers are patient. They start their construction project by hacking out the entrance in the outer growth rings. They then abandon the site, sometimes for years.

In the woodpeckers' absence, fungi take over. They're on the job mere minutes after the first turn of the shovel—or, in this case, the first blow of the beak. There are a multitude of their spores in every cubic foot of air, and they immediately land on the site of the damage. Fresh fungal growth appears and starts to decompose the wood by eating it alive. The wood becomes soft and mushy, so, after years of waiting, our woodpecker couple can finally return to building their home without getting a headache. Once the cavity is ready, the woodpeckers can start a family. Things don't always work out, however. Other birds are quick off the mark and try to move in uninvited. Black woodpeckers can get rid of shy stock doves with a few energetic threats. Jackdaws, in contrast, hang tight and keep possession of the cavity, forcing the woodpeckers to start all over again. Luckily, they usually have a number of living quarters to choose from, partly because males and females prefer to sleep in separate bedrooms.

Over the decades, the tree cavities slowly continue to rot, and the floor sinks. They're too deep for black woodpeckers when the young can no longer reach the opening out of which they must one day fly. And so the retiring stock doves get their turn, after all. They simply raise the level of the floor by adding nesting material (a solution that doesn't occur to the woodpeckers).

The cavity and the entrance hole continue to decompose. The entrance eventually becomes wide enough to admit owls. They also like to use the cavities, which have grown to an impressive size by now, and they often take up residence for many years. One or two little yellow necked mice also make themselves comfortable in the dry warmth inside the tree, losing morsels of food and flakes of skin.

And this is where our picky mealworm enters the picture. It is not until now—after the lineup of cavity squatters has moved in and out as I've just described—that it makes itself at home. The reason is its distinctive food preferences. Mealworm beetles—or, to be more precise, their larvae—love the mixture of fungi-processed crumbly, mealy morsels of wood; insect remains; feather fluff; and flakes of skin, along with all the odds and ends the squatters have let trickle down from above. Bon appétit![7]

It should come as no surprise that populations of this mealworm and similar species are now endangered. Trees that rot for decades in the way I've just described are not particularly prized in commercial forests. They are often cut down and sold the moment the initial woodpecker damage is observed, before internal rot makes the wood less valuable. True, here and there individual trees are left standing to do at least a little bit for species conservation, but such lonely outliers are not much good on their own. You need a large number of these kinds of cavities to safeguard populations of all the living things that are part of this delicately balanced community.

So the beetle faces the same fate as the hoverfly, and there's only one way to support these and other species. Instead of attempting a rescue mission by saving scattered individual trees from being harvested, large areas of forest should be taken out of commercial forest production completely. The claim that well-regulated forestry can do a good job of combining commerce and conservation across the whole forest should be banished to the realm of myth and legend immediately.

Just as trees are not defenseless when bark beetles attack, so too they don't have to stand idly by accepting whatever the climate throws at them. And it's not only that they are capable of enduring an enormous range of temperatures, but also that they can actively influence the weather, as we shall see in the next chapter.

12

What's Climate Got to Do with It?

TREES ARE NOT completely at the mercy of variations in climate, at least not if they come together in large forests and operate as a community. There are limits to what they can do, of course, but if trees work together, they can not only regulate the humidity and air temperature in the forest but also influence other factors for miles around. Recently, a report from an international group of researchers looking into changes in forests because of commercial practices in Europe made me think more about this.[1] The focus of their study was the switch from the deciduous forests of old to coniferous plantations.

What the scientists working with Kim Naudts at the Max Planck Institute for Meteorology were most interested in was how trees reflect light. Deciduous trees are lighter in color than conifers. Plus, ancient beech forests, which once dominated Central Europe, transpire up to 6,800 cubic yards of water per square mile on a hot summer day, which cools the forest air for a long way down below their crowns. The dark green crowns of conifers absorb more solar radiation, which has a warming effect.[2] Conifers are also thriftier with moisture, so the air in coniferous forests is drier, and the way conifers manage water intensifies the warming effect of their dark needles.

The focus of this chapter, however, is not the effect of forestry on climate change but whether there's a reason conifers behave as they do. Whether or not they're being grown commercially, these are not trees bred specifically for plantation life; they behave the same as wild trees growing in ancient forests in cooler climes, which is where they originated.

And this is precisely the part of the world where their behavior could be advantageous. Out on the taiga, summers are short, often lasting only a few weeks. That means the conifers there have little time to grow, let alone form cones and propagate. It might well be that by increasing the ambient temperature, these forest ecosystems are simply trying to extend the warm season by a few days, buying themselves time that could be crucial to their success. That sounds logical, but right now it's pure speculation.

Their strategy for surviving winter is further proof of how desperately spruce and pines need every warm day they can get. Unlike deciduous trees, these conifers keep their narrow, pointed leaves on their branches all the time so that they can get to work immediately when conditions are right. In

Central Europe, this can be as early as the end of February or the beginning of March, when the deciduous beeches and oaks are still deep in their winter slumbers. As soon as the sun warms the air (and the dark crowns of the conifers), spruce and pines begin producing sugar.

That sounds logical, too, and can be seen every year on sunny days as winter wanes. Yet it's only half the story. Another feature of conifers appears to contradict the processes I've just described. Throughout the endless forests of the taiga, there are other substances floating in the air: terpenes emitted by spruce and pines. When fresh, tangy scents greet your nostrils as you're out walking in a coniferous forest, these are the substances you're smelling.

The hotter the sun, the stronger the smell, and this connection is probably not coincidental. Researchers have discovered that water droplets attach themselves to the scent molecules the trees emit. Clouds don't just happen. Water molecules often bump into each other in the air, but instead of sticking together, they part. If that happened all the time, it would hardly ever rain. For precipitation to happen, a large group of water molecules has to clump together and get heavy enough to fall as a raindrop.

These clumps don't form unless there are small particles wafting through the air that the water molecules can adhere to. There are lots of these particles out there in nature: ash from volcanoes, dust from the desert, tiny salt crystals from the ocean, but above all, particles actively emitted by plants. And here our conifers play an important role. They discharge enormous quantities of terpenes into the air. The hotter it is, the more they emit. The terpenes would probably just have a fresh, tangy smell if it weren't for a second component: cosmic rays, which are tiny particles from the universe. They rain down on us constantly, even passing right through us—even through you right now as you're reading this book. These rays make the terpenes ten to a hundred times more effective than they are naturally, because they make the trees' discharges clump together. When terpenes clump, water attaches to them very easily.[3] And so the endless coniferous forests of Siberia and Canada summon, or I should say create, rain all by themselves.

Even when a forest creates clouds that don't drop rain, that's still a plus. The high swirling mists cool the air considerably and slow the rate at which water evaporates from the ground. If the trees manage to conjure into being not just a couple of clouds but a substantial thundercloud, that's like hitting the jackpot. Even a small thundercloud can easily hold 130 million gallons of water.[4]

Now, of course, we have a problem. On the one hand, coniferous forests heat up the air with their dark crowns, which makes them ready to start growing more quickly in spring. On the other, they cool down the air by forming clouds. Is this all just an accident? One of nature's whims? Am I perhaps seeing connections where there aren't any?

Maybe taking a look at the seasons in which these phenomena occur will help. When the first warm days of spring allow spruce and pine to get a jump start on the growing season, it's still relatively cool. Thanks to the dark needles, the sun can make the air a tiny bit warmer, which immediately heats the trees' tissues and helps them kick into gear much earlier than deciduous trees, which must first go through the laborious process of growing new leaves. "A tiny bit warmer" really isn't much here: all it takes is temperatures above 25 degrees Fahrenheit. Then spruce can start producing sugar, but they're not emitting many terpenes.

It would be counterproductive to put up an enormous sun screen made of mist this early in the growing season. At temperatures up to 40 degrees Fahrenheit, the spruce is metabolizing but not increasing its girth, which means that the tree is basically marking time. Production doesn't kick into high gear until temperatures exceed 50 degrees Fahrenheit. At this temperature, sunshine is being converted into sugar, new wood is being grown, and energy is being invested in extending the length of its branches and roots. Therefore, it doesn't make sense to do any cooling until it gets really hot later on in the summer. Conifers begin to suffer significant damage once temperatures exceed 104 degrees Fahrenheit.[5]

Does this sound too warm for Siberia? It gets as cold as it does there because it's so far away from the moderating influence of the ocean. In winter, water in the ocean acts like a heater, and in summer, it acts like an air conditioner, as the air passing over the water is either warmed up or cooled down. In the interior of the continent, this effect is barely noticeable, which is why temperatures far inland are so extreme in both winter and summer. Therefore, it's only logical that the conifers that are so widespread in these regions have developed systems to warm up as well as cool down, and the latter also ensures that, every once in a while, it rains.

When you look at photographs of the taiga or perhaps even visit, you'll notice that spruce and pines are not the only trees in this landscape. By no means. The deciduous faction is well represented, as well—especially by birches. If spruce manage the somewhat adverse climatic conditions remarkably well, then birches must suffer correspondingly badly. Deciduous trees emit far fewer organic substances, and at the onset of spring, there's no dark foliage to warm their cold and clammy little trunks. They get a far later start in spring than conifers. In addition, their leaves need to be completely replaced every year, which costs them extra energy.

What's the advantage to being deciduous? There are two. The first has to do with drought. In winter, deciduous trees lose less water than conifers, because on those few days when it warms up, they don't transpire, because they don't have any green on them. The second has to do with offspring. The seeds of deciduous trees, such as birches, poplars, and willows, travel much farther than seeds from cones and can arrive quickly after forest fires and be the first to form new forests. The older the forests become, the more spruce and pines prevail. Then the forest gets darker, and the light-loving deciduous trees disappear once again.

EVERY TREE HAS its ecological niche and preferred climate, and Central Europe has a few peculiarities that make the lives of these gigantic plants quite a challenge despite the relatively mild temperatures. The climate here is described by the cryptic sequence of letters Cfb. These letters stand for a temperate climate with warm summers and an even distribution of rainfall throughout the year.[6] That sounds good: moderate, warm, and moist. But more important than these three adjectives are the extremes: heat waves above 95 degrees Fahrenheit and cold snaps below 5 degrees Fahrenheit, which are a challenge for native trees.

Below about 23 degrees Fahrenheit, trees contract, which is to say, they get skinnier. That's not because the wood itself shrinks, because a purely mechanical process wouldn't reduce the diameter of the tree by much, and it can lose almost half an inch. What's happening, it turns out, is that water is being drawn farther inside the tree, a process that is reversed on warmer days.[7] Clearly, trees do not shut down completely when they take their winter break.

Even the oak, that champion of extremes, can be pushed to its limits in severe cold. An oak can survive these conditions only if it has grown old without any wounds to its trunk. If it is unscathed, its wood is flawless and evenly structured. Woe to a tree that has had hungry deer taking bites out of its bark or tractor tires disturbing the base of its trunk. If the oak has been damaged, it will have had to seal off its wounds and cover them with new bark. And that's when the problems begin.

Normally, a tree's woody fibers are organized in a uniform vertical pattern to avoid stress in its trunk. When a storm bends the tree over a bit, this vertical arrangement ensures it is flexible enough to sway back and forth. Wounded trees, however, have other priorities—at least around the wound site. They have to grow new bark over the exposed wood, which causes them to call upon the cambium. This crystal-clear layer divides to form new bark cells on the outside and new woody cells on the inside, which is how a tree increases its girth every year so that it can support its growing crown. In its haste to heal, the tree can forget its regular growth pattern in the area around the wound, and thick swellings form under the new bark.

The wood thickens because the tree is rushing to heal itself. If it dallies, fungi and insects will have a better chance of brazenly pushing their way inside. In the chaos, the tree has no time to worry about neatly organized fibers, and at first, that doesn't matter. After a few years (after all, trees are really slow), the task is done. The wound has healed over, though there will always be a thick scar to show where a deer or tractor hurt the tree. But just because the trauma is over doesn't mean it's forgotten. Now it all depends on what other stresses come along. One day, cold temperatures arrive and our veteran tree is at a distinct disadvantage as the damp sapwood inside it freezes solid and the ice threatens to shatter its trunk.

Healing the old wound has resulted in a chaotic bundle of fibers that exert varying amounts of pressure on the wood around them when they freeze. On clear, frosty nights, cracking noises reverberate like rifle shots through the forest. These are not hunters at work but the oaks. Their woody tissue fails around the old wound and splits open so abruptly that the sound travels for miles in a phenomenon known as frost cracking.

IN HOT SUMMERS, other problems arise. Normally trees regulate their microclimate themselves. They all sweat together, as shown by the enormous use of water on hot days. The moist air brings the temperature down by a number of degrees, and the trees maintain the temperature they enjoy. However, if it's dry for months on end, at some point reserves in the ground are used up. The first thirsty trees send out a warning over the wood wide web and advise all the others they'd better be frugal with the last few drops.

If it remains dry and the sun burns hot in the sky, the only thing to do is jettison leaves. At first, just some of the leaves turn yellowish brown and fall to the ground. This rids the trees of some of the surface area that transpires, but it also means sugar production is drastically reduced. Hunger—the lesser of two evils—replaces thirst.

If the rains return in mid- or late summer, it will be too late to grow new leaves. There's time for that only until the end of June. With reduced photosynthesis, the trees must tap into reserves they have set aside to grow new leaves next year, and these reserves will get depleted before spring rolls around. If the trees get attacked by pests, they will have barely enough energy to fend them off. All this is made much more dangerous when heavy machinery used by modern forestry compacts the soil around the trees, which then can't store much water because the pore spaces in the soil have been flattened under tons of weight. For the trees that remain after the forest's water tank is crushed when other trees are harvested, thirst becomes more of a problem in hot summers, and the situation is exacerbated by the greenhouse effect.

Current climate change is heating up tempers as well as the atmosphere. For some, climate change means the end of the human race and all life on the planet; for others, it's a natural phenomenon and the climate has always changed. The latter point of view is clearly true but not very helpful. We all know that ice ages and warm interludes have come and gone over vast periods of time. Even though I believe that human-made climate change is real and already having dramatic effects, I'd like to start by looking at arguments on the opposing side. Let's take a look at the natural cycles of carbon dioxide on an appropriately vast time scale.

In the Cambrian, about 500 million years ago, there were already vertebrates very distantly related to us. They had to deal with levels of carbon dioxide that sound like something from science fiction. Whereas we have pushed the amount from 280 ppm (parts per million) to more than 400 ppm, the amount in the Cambrian was more than 4,000 ppm. Then it sank, before rising extremely again to around 2,000 ppm 250 million years ago. Why didn't the earth collapse from heatstroke?

If we consider the future predicted by many scientists if we get to just a few hundred ppm more than preindustrial levels, life at these levels should be basically impossible. But clearly it was possible, or humans would never have existed. It's a question of the speed at which change happens—and thus the chances species have to adapt—that decides whether climate changes like these are catastrophic or benign.

Basically, the rate of change is slow. Among other things, it's tied to plate tectonics and continental drift. When continents are moving quickly and the African plate, for example, is being rammed under the Eurasian plate, mountains rise where the plates collide. The higher the mountains tower over the land, the faster the rock of which they are made erodes. You can see this in the Alps, where piles of small loose stones cover the feet and lower slopes of the mountains. This scree is weathered down to sand and dust to be washed away by water and deposited elsewhere, along with the carbon dioxide captured in the redistributed material. In times of low tectonic activity there is a correspondingly lower supply of newly eroded rock. This is when volcanoes enter the picture. They spew out molten rock, and the intense heat releases the carbon dioxide bound up in it. In tectonically quiet times, more carbon dioxide is released by volcanic activity than is recaptured in eroded rock. When the earth is pushing the continents together with great force, the situation is reversed.

Does that sound complicated? I think so, too, and yet it's important to understand these vast cycles to get the big picture. If volcanic activity were not releasing carbon dioxide out of rock and returning it to the atmosphere, we'd be facing a completely different problem. At some point, we'd run out of carbon dioxide, and that would be fatal. Oxygen is our most important elixir of life only because we need it for the cells in our body to burn carbon compounds. Without carbon, the purest breath amounts to nothing. Plants fish carbon out of the air around them and store it in the form of sugar and carbohydrates. It is vitally important for us that we don't run out of carbon dioxide.

But that's exactly what the far distant future seems to hold. For hundreds of millions of years, leaving aside the fluctuations, the concentration of carbon dioxide in the atmosphere has been falling. The warmer the world gets, the more this process speeds up, because warmth increases the rate of erosion and therefore the rate at which the gas binds to tiny particles.

"Hundreds of millions of years" are the words to take note of here. Yes, over the very long term, the concentration of carbon dioxide can and probably will sink further, but it won't completely disappear because volcanoes will always be releasing it. And life will adapt to lower levels, as it always has. Much more important are the relatively rapid short-term changes that upset the finely tuned balance, such as the current increase in carbon dioxide concentrations happening because when we burn fossil fuels, we are taking carbon dioxide out of the ground and releasing it into the atmosphere at an unnatural rate. Relatively rapid short-term changes have happened time and again in Earth's history, and every time, many forms of life died out abruptly. At the moment, we're staring at rising carbon dioxide levels like a deer caught in the headlights, and the thing that should concern us most is the rate of change. Higher temperatures are not in and of themselves bad, as long as nature has time to adapt.

The problem is particularly obvious with trees. Populations of trees move very slowly. They can't simply shift 100 miles north every few years, even if the wind or birds carry their seeds for them. When a jay carries a beechnut in that direction, the seed must sprout, grow, and then at some point, when it is a mature tree, produce offspring off its own. Journeys north like this are therefore constantly being interrupted by pauses that last for centuries. And so the average rate of advance is about a quarter mile a year. This means that escaping north takes thousands of years, time that Beeches, Oaks & Co. don't have right now. And the species that are already up in the North need to figure out how they are going to manage as conditions change.

The enormous coniferous forests that can magically summon clouds by emitting terpenes have to exert themselves far more vigorously in these times of climate change. Change is happening particularly quickly in northern latitudes, and the more intensely the sun burns in the sky, the more substances spruce and pine release to produce the cooling clouds. It is really astounding how much these forests have been able to do to help themselves—until now.

It hasn't been possible, of course, for the trees to react to human-caused changes in the short term. They live far too long to be able to do that. You can only have genetic variations over a succession of new generations, and, depending on the species, these opportunities occur only every couple of hundred years—sometimes only after thousands of years—when the mother tree ends her life and makes space for her offspring. And when fluctuations within the life-span of a tree become the norm rather than the exception, then it, or rather the whole forest, has to come up with a strategy to compensate.

Trees have to be able to move from the spot where they are growing, yet not a single one is capable of doing that. That is a real dilemma, because each species is adapted to a particular climate where they can thrive. Whereas coconut palms need tropical temperatures year round and die a miserable death in the cold, deciduous trees can't sustain a growing season without taking a winter break. That's good, you could say. That means each species grows in the very place where climatic conditions suit it to perfection. And it's only because the earth features such a wide array of conditions that tens of thousands of species of deciduous and coniferous trees evolved in the first place.

Now these climatic conditions are changing all the time, and as far as trees are concerned, they are changing relatively fast—even in Europe, where temperatures in recent centuries have fluctuated considerably, above all in the period known as the Little Ice Age. Scientists at the University of Colorado Boulder blame a series of volcanic eruptions for this particular fluctuation.

After 1250, four fiery mountains near the equator erupted and their ash quickly entered the atmosphere and spread across the globe, blocking sunlight. As a result, so the scientists say, temperatures fell and glaciers expanded. The reflective properties of ice intensified the cooling effect, and temperatures sank even further. On average, it became 4.5 degrees Fahrenheit cooler, which is a whole lot when you consider the consequences that a warming of 3.5 degrees is predicted to have today. It wasn't until 1800 that things began to gradually warm up again. This was a very stressful time for trees, because each individual tree had to stay put and stoically endure the changes thrown at it—and it wasn't just cold all the time; some of the summers were extremely hot.[8]

Trees have only two strategies to survive this roller coaster. First, most can survive in a wide range of climates. You can find beeches from Sicily to southern Sweden and birches from

Lapland to Spain. Second, the genetic bandwidth within a species is very wide, so in a forest you can always find individual trees that can deal with the new conditions better than most of the others. In times of upheaval, these are the trees that reproduce and form new stands better adapted to the new normal. But for the extent of the fluctuations we are having today, neither the beech's strategies nor those of the cloud-creating conifers is enough. If it gets too hot, trees will get sick and quickly be killed by bark beetles—weakened spruce and pines are their bread and butter, after all.

With the escape from high temperatures required today, it all comes down to how fast the trees can travel. Does that mean that species with small seeds capable of flight have an advantage? Not necessarily, because trees have a big problem when it comes to getting the next generation started. They have to provide their embryos, the seeds, with an energy reserve in the form of starch or oil and fat. In the first days of its life, the sprouting seed has to grow without being able to make any energy from photosynthesis. Roots penetrate the soil to get water and minerals, while above ground, cotyledons, or seed leaves, develop, which look very different from the later solar arrays that are so distinctive for each species. Only after the leaves have grown can water and carbon dioxide be transformed into sugar using light; only then is the tiny sprout no longer dependent on the energy reserve bequeathed to it by its mother. And the size of this energy reserve differs for every species of tree.

Let's start with the smallest seeds, the seeds of willows and poplars. They are so minuscule that you can just make out two tiny dark dots in the fluffy flight hairs. One of these seeds weighs a mere few millionths of an ounce. With such a meager energy reserve, a seedling can grow only a fraction of an inch tall before it runs out of steam and has to rely on food it makes for itself using its young leaves. That only works in places where there's no competition to threaten the tiny sprouts. Other plants would cast shade that would extinguish the new life immediately. And so, if a fluffy little seed package like this falls in a spruce or beech forest, the seed's life is over before it's even begun. That's why willows and poplars are among the pioneer species that do well settling unoccupied territory.

You find the conditions they enjoy after a volcanic eruption, an earthquake, or a wildfire that completely obliterates plant life. In these landscapes, the tiny seeds can make good use of their advantages. Without rivals, they can grow up to 3 feet tall in their first year—and after that, emerging non-woody plants and grasses can no longer inhibit their growth. The trick, of course, is to be the first to find these places. Because the fluffy seed packages don't have an onboard computer, let alone any steering mechanism, the only way they can do this is by sheer numbers. A few of the vast number of fluffy fliers will land in a suitable spot. A mother tree of one of these pioneer species releases up to 26 million seeds—every year. To keep the species going, all it takes is if every twenty to fifty years one of the little ones gets a good start and reaches an age when it, too, can reproduce. Does that sound wasteful? As trees have no idea where the ideal spots are, playing the numbers game is clearly the only way to get to the places they need to reach.

There are other ways of doing things, however, as the beech and the jay show. Airmail is a good choice if you want to travel. Jays fly little more than half a mile before they deposit their plunder, but that's quite far enough for the beech. Its goal is not to reach those undisturbed spaces devoid of trees that are rare in Central Europe but simply to have the opportunity to travel in the first place. Trees need to be able to constantly expand their range a little to the north or south to follow climates that are always heating up or cooling down, even without the help of humans.

These changes usually happen so slowly that the restricted reach of birds is quite far enough. And for the beeches we're talking about just one option for a small portion of their seeds, most of which will happily fall to the feet of the mother tree to sprout and grow in her shade. Beeches, as well as Douglas firs and other socially oriented species, love their families. If that sounds exaggerated, it's worth taking a moment to listen to the Canadian scientist Dr. Suzanne Simard. She discovered that mother trees can sense through their roots whether the seedlings at their feet are their own children or the offspring of other trees of their own species. They support only their own children, by providing them with sugar through connected root systems—that is to say by suckling them. But that's not all. To help the young trees, the parents step back underground, leaving the little ones more room, water, and nutrients.

Where such close relationships exist, such a strong pull to network with your family, what sense does it make to allow your offspring to be carried far away by the wind and by birds?

Not much, and that's why beechnuts can't fly. Most simply fall down from the branches and land in the soft leaf litter of the mother tree. Fast trips are not their thing.

However, if a beechnut happens to land in a spruce forest—because that's where a jay is setting up its winter stores—the seedling that sprouts has a good chance of surviving. It can handle low light, and it is patient. Fraction of an inch by fraction of an inch, it grows slowly upward until eventually it reaches the canopy, where it can enjoy full sun. Now it produces seeds of its own. Alone like this, far from its family, it must have a harder time than the other beeches, but it is fulfilling an important task. As soon as the temperature shifts a bit, it is the germ of a forest that will spread just a little farther.

Historically this has been a brilliant strategy, but right now trees with large seeds are moving too slowly. Should we help them? Couldn't we export beechnuts to Norway and Sweden and get a jump on establishing new beech forests, creating space for other trees, such as trees from the Mediterranean region (which have the same problem), which we could plant in forests in Central Europe?

Apart from the fact that there are already beech trees in southern Sweden and southern Norway, I don't think this is a good idea. We know too little about how climate change will play out, and we don't know how local climates will develop. Warming doesn't mean that there will never be cold winters. It just means that cold winters will not happen as often. And if we import species of trees that love warmth, they might freeze to death in an exceptionally cold winter. Apart from that, a tree like the beech comes with a whole ecosystem containing thousands of species. Therefore, we'd do better to concentrate our efforts on not allowing temperatures to rise too quickly—then the trees, with their slow rate of travel, will be all right.

There is, however, another kind of heat that can be even more dangerous for trees. And because some species of trees are pretty much like full tanks of gas, the situation can become too hot to handle.

13

It Doesn't Get Any Hotter Than This

A FOREST IS AN enormous storehouse of energy—its biomass, both living and dead, contains a great deal of carbon. Depending on the type of forest it is, it can contain more than 300,000 tons per square mile, and if it burns it releases roughly 1 million tons of carbon dioxide (because of the two oxygen atoms that are added when wood burns). In coniferous forests, the trees also contain dangerous flammable materials: sap and other easily combustible hydrocarbons. No wonder forests are always catching on fire. Huge fires get going that often rage for months. Did nature make a mistake here? Why did evolution create species that are like open gas canisters?

Deciduous trees, after all, show that there are other ways of doing things. As long as they're alive, they're absolutely immune to fire. This is something you can easily test for yourself (but please with just a single green twig). No matter how long you hold a flame underneath it, the twig will not burn. Spruce, Pines & Co., in contrast, ignite easily even when they're fresh. But why?

The opinion among forest ecologists is that in northern latitudes—the place most conifers call home—fire is a natural force for regeneration and even serves to preserve species diversity. Under the headline "Fire Creates Species Diversity," the website Waldwissen.net, which provides information on forestry for federal forestry administrators and professionals, published an article that sings the praises of fire.[1]

I find this an odd idea for many reasons. First of all, we have the idea of species diversity. To make a quantitatively verifiable statement on this topic, you have to know how many species are in the forests in the first place. Many organisms have not yet been discovered—even in Central Europe, a region of the world that has been relatively thoroughly researched. And even if we know a species exists, we've often not done much research into how it lives and how widely it is distributed. Discovering a species exists doesn't say much except that it has been spotted somewhere at least once and we have its description on file.

There is a small beetle that lives in undisturbed forests that has been found in the forest behind the lodge where I live. In the state of Rhineland-Palatinate, it has only ever been seen in two other places, and these sightings both date back to the 1950s. Does this mean that this species is extremely rare? We don't know because, as in many other specialized fields, there's no money for further research. What we do know is that a weevil like the one found in the forest behind my home can survive only when conditions remain unchanged for a very long time. And because conditions in ancient forests rarely change much for hundreds, even thousands, of years, the little beetles have lost their ability to fly. Why roam far afield if life is good close to home?

And so it's no surprise that populations of insects such as this one remain in the same place for a very long time. Their presence indicates that the forest has remained relatively undisturbed for centuries. A forest fire—probably extending over a vast area—would throw the whole system off balance. Where could the tiny inhabitants flee to? And, even more importantly, how quickly could they run? A weevil could hardly escape a mighty wall of fire on foot. No, as far as I'm concerned, everything points to the fact that most forests in their natural state are not acquainted with fire.

There are other reasons I find it odd to categorize forest fires everywhere as inherently natural phenomena. People have been playing with fire for hundreds of thousands of years, and depending on your definition of people, for much longer than that. If you include our forebears such as *Homo erectus* (upright man), then fire has accompanied our ancestors for about 1 million years. This is what researchers reported after they came across what were clearly cooking fires fueled by twigs and grasses in Wonderwerk Cave in South Africa.[2] Examining the remains of teeth led to speculation that this relationship could go back twice as far,[3] and that modern humans developed their large brains because they enjoyed hot meals. Cooked food contains more energy and is easier to chew and digest than raw food. No wonder people and fire became inseparable from that time on.

Fire, therefore, has not been an exclusively natural phenomenon for quite some time. Everywhere our ancestors lived, it was one of the very first by-products of incipient civilization. So how can we distinguish between fire of natural origin and fire started by people? From our vantage point, it is impossible to distinguish between the two in places where there were both people and trees. How can we tell from charred layers today whether a forest fire was started by lightning or by a cave dweller building a fire? You can't conclude that fire is a natural cycle in these places simply because they happened regularly and forests always regenerated afterward. The most you can say is that fire accompanies human settlement.

A strong argument against fire and forest naturally going hand in hand is the existence of individual trees that are extremely old. Take, for example, Old Tjikko, a spruce that stands in the Swedish province of Dalarna. According to scientific analysis, this puny little tree is bent under the weight of surviving for at least 9,550 years, and it could get older yet. If a forest fire had swept through the region in those years, Old Tjikko would have departed for the realm of its ancestors a long time ago.

Yet thousands of square miles of forest burn every year in Europe alone, above all in the south. There are many reasons for this. First, many forests have been cleared. This dates back to the days when Romans were cutting down trees to build their ships. Once the trees were gone, bushes took over, and forests could not be reestablished because cattle, sheep, and goats were pastured there, which meant that no little trees had a chance of growing up. The shrub-covered landscape lay—and still lies to this day—defenseless in the searing heat of the sun, and its dry bushes and grasses offer prime fuel for flames. In recent times, the forests that remained—often consisting of different kinds of oak—have mostly been replaced with pine and eucalyptus plantations. Unlike oaks, both these species burn like tinder, as the forest fire statistics of the last few decades clearly attest.

But the spark that sets off the wall of fire must come from somewhere. In rare cases, lightning is the culprit. But for the most part, it is people who want to make the forests burn, for a variety of reasons. Often, it's because they want places to build, which isn't allowed in forests in Europe. Once the forest is gone, new hotels and homes can spring up. This is what happened after devastating fires in 2007. In Greece alone, more than 580 square miles of forest fell victim to flames, including almost 3 square miles in the Kaiafas Lake preserve. But instead of allowing the area to regenerate naturally, the government decided to allow tourist facilities to be constructed and to retroactively approve about eight hundred structures that had been built there illegally.[4] Perhaps even worse are the motives of some firefighters. Firefighters put their lives on the line every time they go out to make sure people and property are safe, which makes it all the more reprehensible that there are a very few who, to ensure their jobs are secure, start fires themselves when things slow down.

Most fires have one thing in common. They can, directly or indirectly, be traced back to human activity. Even though flaming infernos do not usually have a natural cause, foresters still use them as a burning excuse for clear-cutting. If clearing forests by fire is a natural occurrence, so the argument goes, then removing all the trees at the same time as a harvesting protocol can't be detrimental. After all, nature itself creates open ground.

The opposite is true. Ancient deciduous forests in Europe had one important characteristic in common: long periods without change. And that's why the trees never developed any defense against fire. Despite the fact they are extremely difficult to ignite when they are alive, their skin—the bark—doesn't tolerate heat. Beeches, for example, are so sensitive they get sunburn if they grow in a clearing.

Even though forest fires are rare exceptions in most of the forests around the world, there are some ecosystems that are adapted to such events. Not to the complete incineration of all the trees—that would be an unforeseen catastrophe for any forest—but to fires that burn along the ground. These surface fires are another thing altogether, because they destroy only low-growing vegetation such as grass or nonwoody plants but not the trees—at least, not the old trees. Old trees in fire-adapted ecosystems are outfitted to withstand high temperatures periodically, and you can see this in their bark.

There is, for example, the coast redwood (*Sequoia sempervirens*), one of the mightiest trees in the world. It can grow more than 300 feet tall and live for many thousands of years. Its bark is soft, thick, and slow to burn. If you find one of these in a city park (and you can find them in many city parks all over the world), step right up to it and press your thumb into the bark. You'll be surprised how soft it is. It holds a great deal of trapped air, which insulates the tree most effectively. Thanks to the insulating qualities of its bark, the trunk can survive unscathed a quickly moving front of flames, such as those created by summer grass fires or fires in the undergrowth.

But it is only older individuals that protect themselves this way. Redwood children have such thin bark that they are heavily damaged and often burned up by fire. Redwoods, therefore, expect to face fire over the course of their long lives, but they don't need it to survive—that is where the confusion lies. And, incidentally, that shows that even species that are adapted to fire don't like to burn. Quite the opposite, in fact. In places where fire is a natural component of the ecosystem, mature trees are designed to be difficult to ignite precisely so that large areas are not reduced to smoke and ash.

The ponderosa pine (*Pinus ponderosa*), which is also native to western North America, is another tree that grows thick bark so that heat from forest fires doesn't damage its sensitive cambium. Ponderosa bark works like redwood bark: it protects older trees, and then only as long as flames don't reach the crowns. This is where the needles grow, and they are filled with flammable substances. If the fire gets up there, then it quickly jumps from tree to tree, destroying whole forests. The trees that are supposed to prove that fire is a natural phenomenon show only that even they abhor this element. They have come up with a solution to rare lightning strikes and the surface fires they set only because they have the potential to live for a very long time; these defenses allow them to survive to a ripe old age.

In my opinion, the much vaunted supposed benefits of releasing nutrients by flames and recycling dead biomass through fire are myths that downplay the disruption caused to this sensitive ecosystem by people playing with fire since prehistoric times. In the normal course of events, it is not fire that releases stored nutrients and makes them available to new plant growth in the form of ash; it is the billion-strong army of animal sanitary engineers that undertakes the drudgery of decomposition (and they are completely incinerated in large forest fires, because, unfortunately, the little fellows are thin skinned).

And those that toil in obscurity have a thankless job even in the animal world. The many thousands of species that are small and unattractive are of little interest to humans. Beetle mites, anyone? They make us think of dust mites, and even the thought of them gives us goose bumps. Wood lice? When you find them under the mat at the front door, you don't have much sympathy for them, either. The same goes for many other species that bustle about in the leaf litter under trees. But they are much more important for the ecosystem than, say, large mammals, because without these tiny, overlooked creatures, the forest would suffocate on its own waste.

Beeches, oaks, spruce, and pines produce new growth all the time and have to get rid of the old. The most obvious change happens every fall. The leaves have served their purpose; they are now worn out and riddled with insect damage.

Before the trees bid them adieu, they pump waste products into them. You could say they are taking this opportunity to relieve themselves. Then they grow a layer of weak tissue to separate each leaf from the twig it's growing on, and the leaves tumble to the ground in the next breeze. The rustling leaves that now blanket the ground—and make such a satisfying scrunching sound when you scuffle through them—are basically tree toilet paper.

Whereas deciduous trees drop all their greenery at the same time and stand there stark naked, most conifers keep a few years' growth of needles on their branches and jettison only the oldest. That has to do with their native habitat. In the high north, the growing season is short; there are only a few weeks to grow leaves and drop them again. A tree would barely have time to be green before fall came around and it would have to drop everything again. The tree would be able to photosynthesize for only a few days, and forming new growth or fruit would be almost impossible.

That's why Spruce & Co. retain most of their needles and store antifreeze for the winter instead so that their needles don't freeze when temperatures drop. As soon as the first warm days arrive, the conifers can start producing sugar at full throttle, without having to expend energy and time leafing out. It's as though they are constantly on standby, ready to exploit the brief summer. However, thanks to their larger surface area, they are more likely to get blown over by storms or weighed down by snow. They reduce the risk by narrowing their crowns. Because of the short growing season, they increase their height and girth extremely slowly, and it can take them decades to grow even 10 feet tall. This means that storms get correspondingly little leverage, so the risks and advantages of being green all the time balance each other out.

In climate zones with clearly defined seasons, a tree's greenery must fall, but even in the tropics each leaf eventually serves out its time, and when it is used up and ragged, the tree replaces it with a new one. Inevitably, then, each solar panel, whether it comes from a deciduous tree, an evergreen, or a conifer, drifts down to the ground. And there it would lie forever, buried under a layer of more fallen leaves many tens of feet deep, until one fine day the ground would be depleted of nutrients and the forest would be full to the top with leaves—and then the forest would die.

The billion-strong army of bacteria, fungi, springtails, beetle mites, and beetles is deployed. These tiny creatures are not trying to do the trees any favors. They are, quite simply, hungry. Each one goes about the business of processing its share of the bounty. One savors the thin layers between the leaf veins. The next enjoys the veins themselves. Others concentrate on breaking down the crumbly excrement of those who spearhead the attack.

In Central Europe, this group effort takes three years. After multiple stages of processing, a leaf is transformed into pure fecal matter or, to put it more appetizingly, into humus. Trees can now send their roots out into this layer and use the nutrients that have been released in the decomposition process to build leaves, bark, and wood. Wait a moment.

What happens to the substances that the tiny little guys have eaten and incorporated into their own bodies? Well, the leaves' fate awaits them, too. In the best-case scenario, they are eaten after they are dead and their component parts are excreted. Under less happy circumstances, their end comes more quickly while they are still alive. Small dramas play out in the leaf litter every day. Just as lions on the savanna hunt and eat gazelles, spiders and beetles in the forest hunt and eat springtails. Hundreds of thousands of tiny creatures and many hundreds of hunters are to be found in every few square feet of forest floor that is covered with a thick layer of humus. If you have good eyes and a lot of patience, you can watch the activity for yourself. Depending on the species, springtails can be a quarter of an inch long, and spiders and beetles are even larger.

The substances gathered up in the animals soon get back into circulation when they are excreted, and they are then available to plants as well. There is just one thing the tiny creatures don't like, and that's cold. When it gets too cold, they stop work. And it does get cold in the deeper soil layers 4 to 8 inches below the surface in an intact forest. Humus washed down to these depths by the rain remains basically untouched, even by fungi and bacteria.

Over thousands of years, this blackish, brownish layer increases in depth, and sometimes, because of geological processes, it forms coal. Other material is washed deeper and deeper or, it would be better to say, seeps along with an extremely slow flow of water many layers deeper over the course of decades. And down there, the unhurried underground dwellers are waiting. The deeper they are, the less time seems to matter to them. They, too, prefer organic substances to ash, which brings us back to forest fires. Nature has come up with a much more nuanced, cooler system to cycle nutrients, one in which thousands of species benefit, instead of being incinerated.

These natural recycling systems, however, are mostly no longer functioning as originally intended. People are influencing them and interfering with them in many different ways, and not just with fire.

14

Our Role in Nature

LET'S CUT RIGHT to the chase and start with one of the biggest difficulties here, namely the answer to the question: What is nature? Is it untouched tropical forests or remote mountains with unscaled peaks? What about flower-filled alpine meadows grazed by dun-colored cows with large bells swinging from their necks? Do abandoned strip mines count, where pools have formed and frogs now croak loudly? There are probably as many definitions out there as there are people who love nature. One simple standard definition is that nature is the opposite of culture—that is to say, everything that people have not created or changed. This definition draws hard and fast boundaries around what can be called nature. Other definitions see people and their activities as part of nature. From this perspective, nature and culture cannot be clearly separated.

And this is exactly the problem with the modern conservation movement: What is truly worth protecting and what counts as a threat or even a disturbance? These can be tricky questions to answer definitively when the nature you're thinking of is close to home. However, as soon as you let your gaze wander farther afield, the situation looks quite different. Of course the Amazon rain forest should remain as intact and undisturbed as possible. And Antarctica—a place that under international law doesn't belong to any country—please leave that untouched. You'll find similar attitudes toward other areas, whether coral reefs in Australia or ancient forests in Kamchatka. In your own backyard, a much more malleable rule applies, which states that under some circumstances culturally manipulated landscapes are also worthy of protection, particularly when the original landscape has completely disappeared.

I tend to side with those who argue for a clear separation; otherwise, oil palm plantations in Borneo will one day count as part of nature, too. But how easy is it to make this separation? Which historical epoch signals the divide after which people should be counted as a disruptive influence? If we want to see our species as a disruptor since it arose, what about our predecessors—for example, *Homo erectus*—who differed from us only slightly? There are many questions for which we have no clear answers. I personally draw the line when hunters and gatherers settled down and became farmers. As soon as people did this, selective farming practices began to change species. This is also the time when the landscape began to be intentionally transformed into an ecosystem completely devoted to meeting human needs.

The first irreversible disruptions of the environment become visible, for example, as a result of plowing. When they are drawn over the land, plows disrupt layers deep in the soil profile. The soil retains the imprint of these so-called plow pans for tens of thousands of years. Water drains poorly and even oxygen has a hard time penetrating the barrier they form. As a result, the roots of many species of trees rot when they try to grow down through them, and the trees grow wide, shallow root systems. Then they become unstable, and when they reach a certain height (mostly about 80 feet), the leverage of storms is so great that they topple over.

Just like the birds or bears we've already looked at, we, too, influence the forest and the kinds of trees that grow there, and not only as a result of accidental changes caused by our agricultural practices. Today, 98 percent of the forested areas in Germany are planted, cared for, and harvested on an industrial scale, but even our Stone Age ancestors—who were wandering around not with plows and saws but with bows and arrows—managed to do a fine job of disrupting nature. I'd like to take a look back into the past with you, a few thousand years ago, to see what our ancestors wrought with the modest means at their disposal.

TREES REACT TO changes in climate, and there was a big change at the end of the last ice age. The remnants of glaciers half a mile thick finally melted about twelve thousand years ago, exposing a desolate landscape. There were no forests left in Central Europe. They had all been destroyed as the glaciers slowly advanced from the north. The trees were surrounded, because the glaciers in the Alps also advanced, blocking their path like a gigantic girder across the landscape and preventing them from escaping to the south. Many species died out; others were reduced to a few remnant stands in ice-free side valleys or survived only in the warmer climes of southern Europe.

When the ice melted, the vegetation cautiously returned. At first, there were just mosses, lichens, and grasses, which were quickly joined by miniature bushes and trees. A tundra developed, like the ones found in the northern regions of Canada, Scandinavia, and Russia, where you can still see what a post-ice age landscape looks like. Later, the trees returned. First, conifers such as pines that, along with birches, best withstood the cold that still reigned in these regions. As time went on, oaks and other deciduous trees joined them, pushing the conifers out of most habitats once again.

One representative of the conifer class, however, seems to have dragged its feet: the silver fir (*Abies alba*). It moves very slowly, and so far it has only made it as far as central Germany. You can experience the sequence of the trees' return in the Alps today, by the way. High up on the slopes, where the ice age still dominates, you find glaciers. The farther you descend, the warmer it becomes and the more plants you find—and the plants lower down on the slopes are bigger, too. Four thousand to five thousand years ago, beeches returned from the south to Central Europe. Today, they would form the vast majority of our forests if—and it's a big if—modern humans had not continually interfered by cutting them down and planting other species.

But is it really only modern humans who have done this? After all, our forebears returned to ice-free areas along with the plants, after their ancestors had also been forced by the glaciers to relocate to southern climes. These returnees were far too few to be able to damage the nascent forests. Within the borders of what is Germany today, there were no more than four thousand people roaming around the barren landscape. Then, along with further warming and reforestation, the human population rose, and by 4000 BCE, there were some forty thousand people around. That was just slightly more than one person per 40 square miles. Even if these people needed to burn a lot of fuel, that wouldn't have had much impact on the forest, which grows more than 100,000 cubic yards of new wood annually in an area this size—about the amount of energy used by one thousand modern single-family homes in Germany today.

The problem, then, wouldn't have been that Stone Age people were cold, but it could have been that they were hungry. Stone Age people hunted large herbivores, and large herbivores like to eat young trees. The largest of these herbivores were aurochs, wisent (aka bison), as well as horses and rhinoceros. These species all specialize in eating grass, and they graze grassy plains so thoroughly that they stop any reforestation from happening. This is of critical importance for the discussion we're about to have. If these animals, which naturally shaped their habitat, were present in high enough numbers, then the northern latitudes were probably not forested at all back then.

In the absence of forests, the secret rulers of the ancient landscape were not trees but large herbivores. Herds of grazing aurochs, wisents, wild horses, and deer wandered the grassy plains, demolishing every tree as soon as it emerged. At least, that's the theory. And even if, despite the herds, enough trees managed to get established to form a genuine far-reaching forest, horses and deer would have wasted no time stripping the bark off oaks and beeches, killing them, and the trees' offspring would have been constantly trimmed by hungry herds biting off their buds and new shoots.

It is an undeniable fact that all these large herbivores have disappeared except for the deer. Were they really wiped out by human hunters? Could a few representatives of the species *Homo sapiens* really have had such a powerful impact? Here's where Sander van der Kaars's international research team comes in. The team searched the coastal waters of Australia for traces of excrement left by extinct species of animals. They believe that human hunters who settled the continent about fifty thousand years ago were responsible for the extinctions. They dismissed fluctuations in climate as the cause, because these were not as severe as they were in the Northern Hemisphere at this time. Less than a thousand years after the arrival of the first Australians, 85 percent of the megafauna—that is to say, animals with a body weight of more than 100 pounds—had disappeared.

The disappearance had nothing to do with excessive hunting. Quite the opposite, in fact. In the researchers' opinion, the large animals reproduced so slowly that even a moderate level of hunting inflicted grave damage. The scientists calculated that every hunter removing just a single adult animal every ten years was enough to wipe out the species in a few hundred years.[1]

If large herds of wild cattle, rhinoceros, elephants, and horses really did shape the landscape in Central Europe before hunting humans began to interfere, in a best-case scenario, shrubby growth could have developed but not endless tracts of forest. Of course, supporters of what is known as the megaherbivore theory know that Central Europe used to be almost completely forested. But, in their opinion, this was because of people. Farmers in the Neolithic, they argue, hunted large herbivores intensively and decimated their populations, giving the forest an opportunity nature had not intended it to have—and the forest seized its chance. They support this idea with pollen finds that confirm the presence of grassy plains vegetation prior to this time.[2]

However, there is also evidence of a massive amount of pollen from trees from the same time period. The evidence is not contradictory because even in enormous ancient forests there would always have been areas that were free of trees. These areas could have been swamps, steep slopes, or riparian zones where raging floods didn't give trees the chance to survive for long. The only question is how large the areas of grassland were. Did they dominate or were they merely marginal?

There is one more argument in favor of treeless areas. Aurochs, wisent, and deer are all herd animals. And herd life is only possible on the plains. Have you ever walked with a large group off the trail in a dense forest? Then you'll know that the members of the group spread out and lose contact with one another. You have to keep stopping to wait for stragglers, and because you can't see them, you don't know when they will show up.

For wild cattle, the situation is even more dangerous, because a herd attracts much more attention from predators than single animals do. There are the calls the animals use to communicate with each other; the enormous, strongly scented trail they leave; and, most importantly, the slower speed of the whole group as it has to stop and wait for laggards to catch up. For wolves and bears, it's pretty much an invitation to an all-you-can-eat buffet.

Typical forest animals such as roe deer and their enemy the lynx roam completely alone. It is only at mating time and when they're raising their offspring that you find small family groups of two or three animals. There are also differences in flight behavior. Whereas herd animals often run for miles before they stop, solitary forest animals usually travel less than 300 feet. By then they're hidden in dense undergrowth and can calmly wait and see if the hunter is bothering to pursue them or not.

So we can say that the remains of pollen indicate the presence of areas that were not covered in forest, and the fact that there were large herbivores roaming in herds supports this finding. Human hunting could have led to a drastic decline in their numbers, and the forest could have then reclaimed the now-empty plains. In support of this theory is the fact that most of the large and very large herbivores are now extinct. Mammoths, woolly rhinos, wood elephants and wild horses, aurochs and wisent (except for a few animals in Białowieża National Park in Poland)—none of them exists today. And the warming trend of the past few thousand years is certainly not the only reason for their demise.

So far, so good. But this theory is rather shaky. Let's take a look at the situation from the other side: let's leave the herbivores and look at the trees. Forest trees native to Central Europe, such as oaks and beeches, have completed a long selection process over many generations to become the rulers of ancient forests. A whole range of amazing abilities has allowed them to survive for many millions of years. But there is one thing these trees never developed: protective adaptations against large herbivores. No toxins, no thorns or prickles. Young trees, in particular, have no way of fending off the maws of deer, horses, and cattle. If the megaherbivore theory were true, it would mean that deciduous trees native to Central Europe lived under constant threat with no way to defend themselves.

Okay, we have learned recently that some deciduous trees can identify roe deer and load up with defensive substances when the deer are eating them, but this defense doesn't help much if there is a high density of wild game out there. We know this from the futile efforts of forest owners. Not only are all the small beeches and oaks nibbled on so extensively that the damage stunts their growth for decades as though they'd been bonsaied. But also, when there are too many herbivores around and food is scarce in winter, even chemicals sprayed on the buds to stop them from being munched on are eaten. Deciduous trees appear to be so delicious that beyond a certain density of roe and red deer, there's no way to save them.

These depredations don't happen to typical plains plants such as blackthorn and hawthorn; their names betray their defensive strategies. Even plants such as stinging nettles and thistles have armed themselves. Pointed, hollow needles filled with toxins; bristles that break off easily and remain attached to skin; and tough, bitter fibers are among the methods plants use to keep greedy browsers away. In addition, they can send their seeds out via airmail on the wind or by bird so that they can quickly settle available spaces, even if they're some way away. Beeches and oaks, in contrast, stand there totally defenseless. As I've already described, they drop their heavy seeds directly at the feet of mother trees, and they travel no more than a couple of miles when animals carry them away. It takes these trees thousands of years to move to unforested areas.

The only reliable conclusion we can draw is that there was never a substantial threat from grazing herds. Another point in favor of this interpretation is that it takes a native forest in Central Europe about five hundred years to achieve a stable balance. Millions of hungry hoofed animals would never have allowed the trees so much time. The bottom line is that despite evidence of plains plants and large herbivores, forests must have dominated the area. Even supporters of the megaherbivore theory concede that there were beeches and oaks around. If these had been isolated patches, they would quickly have been stripped bare. And their seeds are so heavy that they would not have traveled hundreds of miles on the wind but only short distances with the help of birds. The fact that these defenseless species of trees were still to be found all over the place speaks against herds of landscape-altering horses and cattle.

This conclusion is bad news for those who misuse the megaherbivore theory for their own ends. Foresters see no problem with clearings regardless of whether they are made by lumberjacks or browsing aurochs; while hunters use feeding stations to boost the number of deer that then devour every small deciduous tree for miles. Both rely on the megaherbivore theory to argue that, based on historical precedent, neither open spaces nor heavy browsing is detrimental to the long-term health of the forest. Hubert Weiger, president of BUND (Friends of the Earth Germany) in Bavaria. has warned: "We are worried that an intellectually interesting and highly technical discussion about conservation ... is being exploited by certain land users... as a political tool to get their damaging objectives implemented."[3]

Apart from natural fluctuations in climate, forests now have to contend with the disruptive climate change we have set in motion. Change is happening quickly—too quickly for trees. In the summer of 2016, I noticed a strange phenomenon when I returned from my summer vacation in Norway at the end of August, and what I saw frightened me. When we traveled to Scandinavia, we had left the forest I manage in a healthy state of green. During our one-week absence, I wasn't concerned. At Hardangerfjord, our destination, it rained so much that I hankered for the weather that was being reported for Hümmel: bright sun and temperatures higher than 86 degrees Fahrenheit. On our return, when we finally caught sight of our native beech forest after a long drive, I didn't feel quite so happy. In the course of those few hot days, many of the crowns had turned brown, and some of the trees were already missing most of their leaves.

I soon convinced myself that it couldn't have anything to do with lack of water. I took a few soil samples from different places and pressed the cores between my thumb and forefinger. The soil didn't crumble; I could press it into little discs that held their shape, which was a sign that there was enough moisture. So what might be the cause?

When trees shed their leaves in summer, it almost always has to do with agonizing thirst. The trees would rather discard their leaves—the surfaces where they lose the most water—before they dry out completely. Unfortunately, that ends the season for them, as they can't photosynthesize anymore. They have enough energy left the following spring to grow new leaves but not for much else. A late frost that freezes the fresh leaves and forces the tree to start over or an insect attack that uses reserves to manufacture defensive compounds—sometimes stresses such as these exhaust beeches and oaks so much that they die. With spruce, death comes in an even more spectacular fashion. Their needles turn a fiery red, and because the dying tree is soon discovered by bark beetles and attacked, not only do the branches lose all their needles but bark also falls off, exposing the trunk.

Back to the summer of 2016. Until August, it had been cool and damp in our area, and trees usually love these conditions.

Usually. In Central European latitudes, too much rain in the summer can favor pests. These had been the cause of an initial leaf drop in July. At this early date, it had been fungi enjoying a feeding frenzy in the leaves, sprinkling them with brown spots or covering them with a thin milky layer of mildew. When their green solar cells were overwhelmed, the trees got rid of them. Some days, leaves were dropping from the crowns as though it were already fall. And then came the quick, unhealthy change to extremely hot, dry weather. Such a sudden switch is enough to throw even the strongest trees off balance. Within a few days, many of the deciduous trees had turned brown and finally discarded leaves the fungi had left undamaged.

It was noteworthy that in the managed stands, where trees are repeatedly felled, the symptoms were particularly prevalent. It's not surprising, because here, in contrast to forests left to their own devices, there are many gaps in the canopy through which the sun can shine unchecked. This means that everything heats up more quickly, the air dries out just as quickly, and all the conditions change much more abruptly. In contrast, in forests that are left alone to regulate their microclimate for themselves, conditions remain somewhat more bearable. In addition, the trees in these forests mutually support each other through their root and fungal networks so that weakened friends can be saved.

And what about weather conditions at other times of the year? As a forester, I've always got one eye on the weather. When it's stormy in winter, I worry about the old spruce that might fall. If they fall, the small beeches under them, which still need the shade of their guardians (even though they are not related), will be exposed to the sun the following summer with no way to protect themselves. If it rains too much, there's a greater risk that the ground will soften and not offer the roots much in the way of support. I prefer freezing cold days in winter, because that means there won't be any precipitation. It only gets really cold under a high-pressure system, when cloudless night skies allow warmth from the earth to radiate into outer space.

And if there isn't any rain or snow, why isn't that a good thing, either? In Central Europe, trees don't get enough water from rain clouds in summer, so they have to draw on reserves that have built up in the ground over the winter. A lot of moisture is stored in the ground when trees are not actively growing, and trees can use this moisture to supplement the rain that falls in the warm months of the year—as long as there was enough precipitation the previous winter.

Hot summer days also worry me. Too many in succession, and the ground dries out and trees suffer. They become more susceptible to disease, as I have already explained. If rain comes, it's often accompanied by a thunderstorm. Right before the storm hits, the wind freshens to storm strength and the deciduous trees that I love so much are now particularly at risk, because they present a large surface for the wind to attack. In winter, which is the usual time for storms in Europe, they present a streamlined, leafless profile, as evolution has taught them to do. So I don't like thunderstorms.

Did you notice? The weather gods can't satisfy a forester like me. By way of apology, I'd like to say that I am simply worried about the trees and their future. Because I pay attention every day, I notice changes that are slowly increasing year by year. It's not just the mild winters, which are being discussed all over the media. There's also a noticeable shift in the seasons. The first snow often doesn't fall until January, even though the forest I manage, which stands at an elevation of 1,650 feet, should normally get at least one sprinkling of the white stuff by November, at the latest. March often slips by without offering any days warm enough for me to sit outside.

The bees are being left out in the cold, because either meadow flowers and other early sources of nectar are late appearing or low temperatures prevent the insects from going out on foraging flights. And while garden centers are offering a wealth of flowers for window boxes and flower beds, we have to wait until the middle of May to fill the garden outside our forest lodge with color. The last snow of the season falls in April and the last frost is sometimes delayed until the beginning of June—so an overly eager purchase of petunias has to be repeated. In Germany lately, it hasn't been getting really hot until August, and in 2016, it didn't heat up until the middle of September. From a meteorological perspective, fall should have arrived by then—with a few final warm days to delight us, to be sure, but temperatures should have been distinctly cooler, especially at night.

All things being equal, we might be fine with everything simply shifting a bit later, but unfortunately, trees' internal clocks are set differently—or perhaps they are simply more stubborn. They are as acutely aware as we are that the days are getting shorter, and they slowly prepare for winter dormancy. Simply holding on to their leaves for another four weeks isn't an option for them, however, because despite the warming trend, they still have to reckon with the possibility of an early onset of winter accompanied by a heavy snowfall, which would punish trees that keep their leaves hanging for too long in the fall sunshine. Their branches would snap and some trees would lose their balance and topple over, which is what happened in a big snowstorm in Germany in October 2015.

The only recourse trees have is to escape north, which is what they are in the process of doing. Or what they are trying to do. It never occurred to us that trees might migrate. The parcels of forestland that we have set aside to mark property ownership create inflexible obstacles for trees wanting to extend their range to cooler climes.

A simple example here is our lawn. When I mow, I always see small oak seedlings poking up out of the grass, which then unfortunately fall victim to my mower. Okay, so the mother oak stands just 100 feet away, but even so, this is a migration, even if it's a very slow one. I've already explained how far birds and the wind can transport seeds, but if we have other plans for every patch of ground where seeds might land, the trees can't even get started on their journey north.

With animal migration, there are international efforts to keep corridors open so that enormous herds of gnus, zebras, and elephants can move from one national park to the next. Even in Central Europe, there is support for animal migration—for example, to help wild cats. Conservation organizations like BUND strive to maintain corridors so that the mini-tigers can expand their range once again and roam everywhere in Germany.[4]

And what about trees? They advance so sedately that tree migration is not on anyone's radar. Even foresters say that Beeches & Co. are too slow to escape to higher latitudes as the climate changes. But the problem here is not that they are too slow, but that we are keeping their populations confined. Every time one of their seeds sprouts somewhere we've not designated as an area for trees, we immediately remove it. Spruce have to grow in district X, and beeches have to grow in district Y. A parcel of land over there is permitted for agriculture; another is listed as a meadow. These rigid boundaries get in the way of what nature has in mind: change.

And this brings us back to my lawn, and, yes, I, too, am guilty. If we've forced our environment into a straitjacket, how can we have any idea how it reacts to climate change? Are trees in Central Europe really too slow to set out for the cooler climes of the North?

Apart from protecting the climate in general by conserving energy, I see designating many more protected areas as a solution. We need areas of wild forests to be like the steppingstones we use to cross water without getting our feet wet. If there were enough of them, wild species could travel freely through our culturally manipulated landscape from one preserve to the next. And if these areas were not too far apart, then perhaps we really could see how trees react to climate change—and we might find out that they don't want to go north at all.

We already know that as long as beech forests are not disturbed by commercial forest practices, they can cool themselves in hot summers. It is only when trees are felled and sunlight penetrates the shade under the remaining dark trunks, and the air down there dries out and heats up, that the giants start to have problems. That makes the solution both clear and simple. Less use of wood - less use of energy = less climate change = healthy, resilient forests. If this could happen on at least a portion of the land, there's hope for the slow giants of the plant kingdom.

SOME EFFECTS OF human activity on nature are much more subtle and more difficult to track than felling trees, simply because cause and effect are much farther apart.

In 1997, I traveled through the southwestern United States with my family for the first time, and I made a return visit twenty years later. We were blown away by the United States. The national parks with their imposing sandstone rock formations are simply breathtaking. Apart from the plants and animals in that vast landscape devoid of people, it was the bizarre shapes of the rocks in Utah that grabbed our attention.

Arches National Park is named for the unusually high number of impressive rock arches that it contains. Some of these gigantic arches look so fragile despite their imposing size that amazed visitors wonder how they have stayed standing under wind and weather for thousands of years. Today, for many of them, this question no longer needs to be asked. Forty-three arches have tumbled since 1977 just in Canyonlands National Park. A number of these tragedies—for that's what they are, not only for tourists but even more so for the Indigenous peoples for whom the arches are sacred—can be traced back to human activity.

A research team at the University of Utah has discovered that the rocks sway ever so slightly for a variety of reasons. Most of the movements are caused by natural events. After earthquakes, temperature fluctuations are the main culprits. The rock expands in the heat of the day and then contracts as it cools down at night, which makes the arches sink slightly.

To get to the bottom of other causes, scientists wired up Rainbow Bridge, which is considered to be one of the highest natural bridges in the world and is a sacred site for the Navajo Nation. Tourists are not allowed near it. Those who want to see it have to travel by boat along a side arm of Lake Powell, and from there a ranger walks them to the observation area. These precautions have less to do with protecting the arch and more to do with respecting the Indigenous groups who live there. Tourism is also not the main threat to the bridge.

As Jeffrey R. Moore's team discovered, the repercussions of human activity are detectable in the rock—every few seconds. The rhythm of waves gently hitting the shore of Lake Powell can be measured on Rainbow Bridge many miles away, where the wave action causes small but continuously repeating vibrations in the rock.[5] If something like that is measurable, it's no surprise that shock waves from drilling in Oklahoma—a distant 1,000 miles away—were also detected. Ultimately, it's difficult to say exactly what has led to arch collapses in the recent past; however, this is a good example of how far-reaching the effects of human activity can be on ecosystems.

Groundwater comes up once again in this discussion. The problem I've just explained about the falling arches gave me an idea. It's pure speculation because, as far as I know, it's never been tested. Water deep underground contains gas. This gas includes the oxygen groundwater crustaceans and other tiny creatures need to breathe, as well as, it then follows, the carbon dioxide they exhale. You know what happens when you shake a bottle of carbonated water: the carbon dioxide bubbles out and the water then contains less of this gas and becomes less acidic.

You could compare the underground water system to an enormous bottle that is constantly being shaken by artificially induced tremors. Surely, that must result in changes in the gas and acid content of the water. That could be the case, at least, in the area around fracking sites where pressurized liquid is used to fracture the ground up to 10,000 feet below the surface. This process produces numerous earthquakes. As collateral damage, many chemicals remain in the ground, their fine particles dispersing and infiltrating cracks in the layers that are being worked. What would the blind crustaceans have to say about all these changes to their ecosystem, I wonder?

IN CENTRAL EUROPE, at least, the majority of the underground flows in this wonderful ecosystem are still untouched, but there have been dramatic changes close to urban developments. For one thing, pollutants from agriculture and industry are seeping underground. For another, enormous quantities of water are being pumped out of the ground every day. In Germany alone, about 13 million cubic yards of water pour out of faucets daily. And then there are industrial uses such as open-pit mines, which are emptied of the groundwater that flows into them in unimaginable amounts. At the soft coal open-pit mine near Cologne, 720 million cubic yards of water was pumped out just in 2004. That's one and a half times as much as all the drinking water used in all of Germany in a year. An underground area of at least 1,000 square miles is affected, where life-forms that we haven't studied yet and whose influence on the natural cycles of life we don't yet know are wriggling around in every cubic yard.

Despite this, there are still large regions where the groundwater remains intact, and together with the deep layers of soil, these really are the last untouched habitats in Central Europe. You don't have to go far to find true nature: it's closer than your closest national park or conservation area but still beyond your reach.

What are both close by and immediately accessible, however, are the results of the last 100,000 years of human evolution.

15

The Stranger in Our Genes

HOMO SAPIENS HAS turned out to be a very successful species (thus far), which is probably an indication of our aggressive nature (I will explain exactly why in the course of this chapter). I don't mean our urge to attack each other but our tendency to attack other species. And this aggression has something to do with our evolutionary success, which has made us what we are today. Perhaps we've been too successful, as evidenced by the decline of other species. Is the desire to disrupt the giant mechanism of nature stored somewhere in our genes? Or have we succeeded in removing ourselves from her clockwork and entered into some kind of parallel ecosystem?

I often hear people talking about how modern humans have halted evolution in its tracks. This notion is linked to medical advances. How many of us would still be alive without appendix operations, insulin injections, beta-blockers, or, even, eyeglasses? Ten thousand years ago, the defects that plague us today would have made us an easy target for predators. The hard truth is that evolution would have weeded us out. So if we survive with medical help despite our physical weaknesses, and if we then pass our defects on to the next generation, does that mean that our species is becoming more fragile and would die out if medical help were to suddenly disappear?

To take a closer look at this issue, we first have to differentiate between two separate questions. First, whether evolution has actually been switched off, and second, whether the use of medical aids might be part of evolution and might indicate an advanced stage of development.

The answer to the first question is clear. Evolution is obviously still hard at work, particularly when it comes to disease. For many people, the level of danger and the pressure exerted on them by disease in their environment are as intense as they have ever been. According to the World Health Organization, in 2015 alone, 200 million people fell sick with malaria, and 440,000 of them died from the disease. In places where there is a great deal of malaria, a rare genetic blood disease is also prevalent: sickle cell anemia. In sickle cell anemia, red blood cells, which are normally round like a disc, become sickle shaped. People who suffer from sickle cell anemia have difficulty getting enough oxygen to their organs, and they often die before the age of thirty. Most people who carry the genes for the disease, however, develop only a mild case, which means that although they have sickle-shaped blood cells, they also have enough normal-shaped cells to lead almost normal lives.

The critical point is the presence of malaria. In this illness, parasites passed on through mosquito bites attack and destroy red blood cells. Malaria advances in stages. Periods of fever, triggered by blood cells bursting en masse, often lead to the complete breakdown of the organism. Carriers of sickle-cell anemia have a natural resistance to malaria. How this works has yet to be adequately explained. In any event, people who have sickle-cell anemia and are significantly affected by this disease have a distinct advantage over those who don't. This advantage means that in areas where malaria is widespread, you also find many people with this genetic mutation.

So the impression that evolution has almost ceased and that humans have achieved the pinnacle of their success is false. It's just that people in wealthy industrialized countries have become somewhat removed from the processes going on around them. But nature is still applying pressure. Cancer, heart attacks, and strokes are only some of the factors we cannot control despite medical advances. Strictly speaking, it's modern civilization that makes modern medicine necessary in the first place. The afflictions that are so aptly named diseases of civilization barely existed thousands of years ago. Dental braces, back operations, and heart bypass surgery are only necessary because of unhealthy Westernized lifestyles. Seen in this light, medical discoveries that have supposedly halted the roller coaster of evolution have simply pushed it in a different direction. Instead of plagues, Cholesterol & Co. are now the factors sorting out the gene pool.

Apart from that, numerous construction sites in our bodies are proof of archaic developmental processes that still function as they used to. Our mouths are losing unnecessary teeth (wisdom teeth), our guts are losing unnecessary appendages (the appendix), and our bodies—to the intense regret of many men—are losing their hair. It's unlikely that people in fifty thousand years' time will look exactly the same as people today. So evolution is going along its merry way, even if we are under the impression that we've reached the end of a long journey. It's just that change happens so slowly that we don't notice.

To make comparisons, it's useful to take a look at the surface of our planet. The appearance of the landmasses, the shape of the continents, seems to be unchanging, even though in school we all learned about the movement of the tectonic plates that form the earth's crust. These plates, which span whole continents, drift on molten rock either toward each other (which thrusts up mountains) or away from each other (which opens up rifts where lava gushes to the surface). North America and Europe, which lie on different plates, are drifting farther apart by about three-quarters of an inch a year, about twice as fast as your fingernails are growing. Apart from a handful of scientists, hardly anyone takes any notice, but over 10 million years—which is just a brief moment geologically speaking—that amounts to 125 miles. When things occasionally get stuck and the jammed plates rip free again, earthquakes alert us to the upheaval.

This raises an important question. Does evolution move at different speeds and in different directions in different places? Whereas some people feel the full impact of the selection process in the form of disease, for others—mostly in industrialized countries—the impact is softened considerably by a wealth of resources. However, what might seem like an advantage for individuals could, in the long term, be a disadvantage for the population as a whole in any given area. Victory over plagues has shut down one of the most important factors that has constantly changed our genetic makeup. If evolution has indeed almost come to a standstill for people in wealthy countries, over many thousands of years, they could be overtaken, genetically speaking, by people who live in countries with less access to medical advances and Westernized lifestyles.

In our modern world, however, such developments are impossible, because our enormous mobility interrupts this process. Modern migrations increasingly blur local differences. Many people today have ancestors who lived in other countries, and this means that a gradual genetic separation of the people of the earth is no longer possible, and the development of different human species is out of the question—at least for now. For that you'd need populations to be isolated for a long time, something that cannot happen in an age of world travel and immigration. Researchers tell us that every person alive today can be traced back to one mitochondrial Eve, who is said to have lived 200,000 to 300,000 years ago. The variations in skin color and other characteristics that have developed since then are disappearing increasingly quickly. What some people mourn as a loss of diversity, others embrace as an opportunity for humanity to bid goodbye to racial differences.

EVOLUTION, HOWEVER, CAN advance in totally unexpected directions. There was a time when *Homo sapiens* was not the only hominin species to walk this planet. To explore this further, let's turn to some dusty distant relatives from the Neander Valley in Germany. Heavily muscled Neanderthals from the Stone Age had a brain comparable in size to ours. Their culture was relatively advanced. There was a division of labor in their settlements. They crafted ornate stone spearpoints attached to wooden shafts, painted their bodies, buried their dead, and spoke in a language that has long since fallen silent.

Scientists assume that *Homo sapiens* and Neanderthals lived side by side in Europe for a few thousand years. Modern humans arriving later on the scene certainly learned something from their stockier neighbors. Could this species of human perhaps even have been mentally equal to the *Homo sapiens* of the time? Scientists debate this question, but in my opinion they do so in a way that is not entirely aboveboard, because earlier *Homo sapiens* differed from people today—in no way at all. And so if you were to answer in the affirmative, what you would be saying is that the intellectual prize we thought was ours alone has to be shared with another species—and that evolution handed the crown to humans not because of the size of their brains but because of their more aggressive tendencies—after all, we supplanted the Neanderthals and might even have used them as a source of meat.[1]

There are some arguments against this interpretation, but an objective discussion isn't yet possible, because people are still attributing to Neanderthals the minimal intellectual capacity the findings allow. Take language, for example. Neanderthals had a small bone under the tongue, the hyoid. This bone is a necessary requirement for speech. There is also a particular gene, FOXP2, that is considered indispensable for understanding verbal utterances, and Neanderthals had this, too. But for scientists, this isn't proof that Neanderthals used language; all it proves is that they were physically capable of doing so. If you take that tack, you could take the presence of eyeball sockets in unearthed Neanderthal skulls simply as evidence that they had eyes. Whether they could actually see with them can never be said with absolute certainty.

The size of the Neanderthal brain has been explained away as either an adaptation to the cold or to their slightly heavier bodies. (The same-sized brain in a larger body means that proportionally the brain is smaller.) However, there are people alive today whose weight and musculature are pushing the Neanderthal range and yet their brains are "only" Neanderthal sized. If that argument held water, body builders would be packing on brain cells as well as muscle while they were at the gym.

A further tenet of science fell by the wayside a few years ago. It stated that Neanderthals and modern humans did not consort with one another; therefore, there's no material from our coarser cousins to be found in our genes. Yet mapping the human genome turned up a number of surprises that caused at least a glimpse of Neanderthal to show itself again. Today, researchers believe that 1.5 to 4 percent of the genetic heritage of most people of European and Asian ancestry comes, somehow, from Neanderthals.[2]

Our extinct relatives greet us through the skin and eye color of many of our contemporaries. A light complexion and blue irises—current scientific opinion holds that these are Neanderthal adaptations to their northern habitat. Here, the sun is less intense, so an inbuilt sun protection isn't necessary. When *Homo sapiens* arriving from Africa had sex with their northern neighbors, they permanently passed these features on to their offspring. Other characteristics from these dalliances are still active today, including the tendency toward depression and addiction to tobacco products.[3]

It was a two-way street, and our genes also entered Neanderthals, something that scientists also rejected as impossible for a long time. About 100,000 years ago, modern humans and their extinct cousins met and became intimate. So intimate that traces of these romantic rendezvous have been found in Neanderthal bones discovered in the Altai Mountains.[4]

Research into Neanderthals is telling. We only attribute to this species of human characteristics that are beyond dispute according to the current state of the research. Wouldn't it be more honest to say that although we know some things for certain, there are others about which we don't (yet) know enough? It seems suspiciously as though we are saying that there just can't be any other beings as intelligent as we are. And nothing is allowed to upset this belief. Not because it is forbidden, but because our instincts struggle against it with a resounding "Never!"

Nature knows of only two paths for the future of every species: adapt or die out. And these adaptations can include changes in intellectual capacity (and that's what we're talking about here). Just to clarify: evolution means adapting to change, not necessarily development in the sense of improvements or even larger brains.

Researchers from the United States suspect that there are definite disadvantages to our powerful brain. They compared the self-destructive programming of human cells with a similar program run by ape cells. This program destroys and dismantles old and defective cells. Their comparison showed that the cleanup mechanism is a lot more effective in apes than it is in people, and the researchers believe that the reduced rate at which cells are broken down in people allows for larger brain growth and a higher rate of connections between cells. This improvement in intelligence probably comes at a high price, because the self-cleansing mechanism also gets rids of cancer cells.[5] Whereas apes hardly ever get cancer, this disease is one of the top causes of death in people. Is the price for our intellectual capacities too high? If our current level of intelligence is not suited to the survival of humankind, it must either be increased or lowered. The latter is probably unacceptable thanks to our ideas about self-worth.

But if you leave aside the amazing things our large brains allow us to do, we might ask whether the amount of intellectual ability that we have today is really necessary for our personal quality of life. What is really important? Of course, there's happiness, love, and security, along with the small daily highlights such as delicious food; a warm, dry home; and other comforts. Do you notice something? All these things have to do with feelings, instincts, and not with intellectual achievements. People living in the year 50,000 CE will be able to live fulfilling lives independent of the volume of their brains as long as they can adapt to constant change in their environment. And they will manage that. No one can escape the network of nature.

16

The Old Clock

NATURE IS CONSIDERABLY more complicated than the finely calibrated movement of a clock, but I would still like to return to the example I gave in the introduction. We've already seen multiple examples of what happens if we carelessly remove one little cog. Just as in the internal mechanisms of the timepiece, this loss triggers a chain reaction that changes the whole system.

But what does it look like if the clock breaks, and we want to restore it to working order? We know nature can heal itself in certain circumstances, but we also know that this takes time. In cases where natural processes take hundreds or thousands of years, could humans step in to speed things up? Particularly when we can see the fruits of our labors—and that's often what it's all about. We want to experience the improvements for ourselves. Why abstain from using vehicles that run on fossil fuels or avoid materials made from them when our great-great-grandchildren are the ones that will see the results of our efforts? And so we step in, determined to achieve positive change as quickly as possible. But when we take it upon ourselves to repair the internal workings of the environment, a big problem surfaces: How do we know when it's broken?

The wood grouse, or western capercaillie, is a good example of one of these attempts to put things right. This large chicken-like bird (depending on the sex, it can weigh up to 9 pounds) lives in boreal coniferous forests—that is to say, it is at home in the spruce and pine forests of the North. There, it eats insects, but mostly it eats the leaves and berries of blueberries. My family and I came across these little bushes everywhere when we were out and about in forests in Lapland. And we saw capercaillies everywhere, too, when we walked in the *fjäll* (mountains). Of course, we were excited every time one of these birds crossed our path, even though they aren't anything special in northern Scandinavia. They're considered wild game there, and they often end up in the stewpot.

It is completely different in Central Europe, where the birds are strictly protected. There's a relatively small amount of habitat suitable for capercaillies here, because sufficiently large natural expanses of coniferous forest carpeted with blueberry bushes are found only in alpine regions. From the perspective of climate, alpine areas in Germany are tiny fragments of Scandinavia. High up in the mountains, winters are long and harsh, making it too cold for deciduous trees. So a handful of capercaillies live here just below the tree line. Of course, tiny scattered populations of any species are inherently unstable. If just a few of them die, the local population will not survive.

In the Middle Ages, the situation in Central Europe was much better for the birds. Forest clearances created half-open landscapes where blueberry bushes grew in abundance. Even today, you can find small blueberry bushes in many planted coniferous forests, particularly in plantations of pines. As they're shaded by trees, the bushes rarely have any berries on them, but they are a reminder of earlier times when the practice of felling trees created the clearings where they love to grow.

This human activity suited the capercaillies as well. They extended their range southward and settled in habitats where they did not originally occur. Forests changed again when modern forestry started. Pastures and agricultural land were reforested, and ravaged forests recovered and filled in. Deciduous trees returned full of vim and vigor to replace some of the gloomy coniferous plantations, and it was much darker below their leafy branches than under pines. Things were not looking good for blueberries and other bushes, nor for red wood ants, which no longer had access to the discarded needles they need to build their nests or the shafts of sunlight they need to warm up enough to go about their business.

The rebirth of beech forests, the native vegetation of Germany, was the death knell for those followers of culturally manipulated landscapes: the capercaillie and the blueberry. Is that a bad thing? No, it's not. All it means is that those species are being pushed back to the places they came from, and, in return, the rare inhabitants of German beech forests are getting their original habitats back.

You could say that everything is slowly recalibrating. You could. But now government and private conservation groups are getting involved. And we're back to the immense clockwork of nature. Is it really broken? Is there something that needs to be repaired? Unfortunately, this question is not even asked, at least not when it comes to the big picture. Instead, the capercaillie has been declared particularly worthy of conservation in the Black Forest, which was originally ancient deciduous forestland. With great fanfare, clearings have been made, and here and there the forest has even been burned to create open spaces for blueberries to grow. The fact that Germany's native forest dwellers are now suffering—ground beetles, for example, which love the dark—is conveniently ignored.

It is a similar story for a small relative: the hazel grouse. Even finding feathers near a construction site means work has to be halted immediately until the situation can be studied in depth. Hazel grouse are in danger of disappearing in Germany. The Eifel originally contained nothing but ancient deciduous forests. The small hazel grouse would never have managed to make its livelihood here if it hadn't been for people settling and clearing land, and creating large juniper heaths with their herds of cattle. In these new, lightly treed habitats—similar to forests in northern Sweden—the hazel grouse felt right at home. Unfortunately for the birds, forests here are also recovering and shading out juniper heathlands.

So a variety of different threads are coming together.

Conservationists who desperately want to help the grouse are pleading for the Eifel to be designated a protected habitat: which means more thinning of trees, which would mean more light reaching the ground and more bushes, so the basic food for this grouse would recover.

Forest agencies are offering to step in and help. Isn't reviving coppicing as a management technique the right thing to do? Coppicing is an old way of managing forests that arose hundreds of years ago out of pure necessity. Timber was becoming increasingly scarce, because it was being used so heavily as fuel and building material, and people were not giving trees time to grow old. Oaks and beeches were being cut down at the tender age of 20 to 40 (instead of 160 or 200), because people just couldn't wait any longer. Acres of forest were razed. New shoots grew from the stumps, and the spindly growth was harvested just a few decades later.

So much of the forest was plundered that it began to look like a carpet full of holes. Hazel grouse love these conditions—it was as though people were doing them a favor. However, more reasonable ideas about forestry prevailed, and strict laws forbade coppicing. At least, they did until the modern need for timber, fired by the boom in bioenergy, set in. And so it was that the new clear-cuts came to be celebrated as reinvigorating a historical forest practice while also helping the small hazel grouse.[1]

Romantic timber harvests and conservation combined? No. Now as then, these are nothing but brutal clear-cuts using automated harvesting machines that weigh many tons. This is no way to establish a proper forest, and whether the hazel grouse driving this process really enjoys the newly cleared habitats mostly remains to be seen. Meanwhile, things are not looking promising for forest species that really belong here, such as the black woodpecker and that mealworm beetle I mentioned earlier.

Keeping meadows open is another example. Meadows provide habitat for a large number of grasses and nonwoody plants. In summer, they are filled with colorful blooms and aflutter with gaily painted butterflies. This splendor attracts many different species of birds, which settle here in great numbers. As agriculture becomes more intensive, this diversity is threatened. Thanks to the rise in the price of corn in response to the explosion in demand for raw materials for the biogas industry, every scrap of spare land is being plowed under and converted to an agricultural desert devoted to a single crop. And there in the meadows, where the idyll still seems to exist, the forest is preparing to reclaim the last isolated valleys and the wetlands along the banks of streams.

Grassy landscapes are under siege. But instead of pointing the finger at agriculture, we are pitting grass against forest, which means that to preserve species that love meadows, it is forests and not fields that have to give way. The methods used to combat the forests mostly look extremely peaceful. There are, for example, the Heck cattle I mentioned earlier. These are supposedly back crosses to aurochs, our original wild cattle, which once grazed the moist meadows alongside rivers and streams. Unfortunately, it isn't possible to breed these extinct species back to life, even if Heck cattle do bear a passing resemblance to ancient aurochs.

When you get right down to it, Heck cattle are nothing but domesticated cattle designed to look like aurochs. This does have one advantage: if you let these cattle graze alongside streams, it looks as though all is right with the world. In reality, though, what this form of farming does is reinforce a widespread misconception, because plains (and grassy landscapes are plains) do not belong to the natural ecosystem in Germany, where all we had was forest punctuated by the occasional mountain range or swamp. The many colorful plants with their butterflies arrived on the coattails of culture and only got established when our ancestors cut down the trees.

There's a simple reason these treeless landscapes delight us so much. We are, from a biological perspective, animals of the plains, and we feel secure in landscapes with extensive views where we can move around easily. Do you remember that megaherbivore theory I mentioned earlier? It is also used here in the confusion between conservation and aesthetics to push the pendulum in favor of the latter. If we were to leave nature alone, then wetland forests would naturally regenerate along our streams and rivers. They don't support colorful plants and butterflies, but they do provide an important habitat for tens of thousands of other species. Think of the tree sap hoverfly. No one even knew it existed until recently. If Heck cattle had created grassy landscapes by preventing the growth of moisture-loving trees, then this fly would have disappeared and we would have been none the wiser. We don't really understand how the clockwork of nature functions, and as long as we don't, we shouldn't try to fix it.

I want to state something clearly here. I don't object in every case to making a special effort to help individual species, even if the species is here as a result of human interference, as is the case with the hazel grouse and the capercaillie. If the species arrived in Germany in historical times and is now threatened globally with extinction, then (and only then) should we go out of our way to help it, even if that means upsetting parts of the native forest ecosystem. However, if there is no global threat, then any interference in the complex web of nature is out of the question.

The red kite is a case in point. This raptor with its imposing 6-foot wingspan is the perfect example of a species that has benefited from cultural changes in the landscape, and it would certainly have been rare in the original ancient forests of Central Europe. It needs open landscapes so that it can glide through the air to hunt for small mammals, birds, or even insects. People, with their desire to clear forests, suited these birds just fine. The two-legged predators created a plains environment with magnificent hunting possibilities.

Every summer out in the fields you can see how adaptable the red kite is. The moment a farmer starts cutting hay with his tractor, there will often be a red kite following along, looking for mice or fawns that have been minced or run over. Most of the global population of about 25,000 to 30,000 individuals lives in Germany. In most other places, red kites have declined drastically. If we now returned exclusively to our native vegetation, it would all be over for the majority of these birds. They have found a second home with us, and their populations are relatively healthy and should therefore be encouraged so that they remain so in the future. The best way to do that is by not only protecting a landscape with small farms and small fields but also by preserving the trees where they nest by establishing buffer zones where there is no active forestry.

Just to remind you: we are talking here of intentional interference with natural processes. We interfere unintentionally all the time everywhere, and here I'd like to restrict my examples to the open countryside. In most places in Germany, we have driven out the ancestral plants (trees) and replaced them with grains, potatoes, and vegetables. What all cultivated species have in common is that they are not native to the area. Even in what forest remains, most of the parcels are filled with nonnative species. Wouldn't it be nice if, at least in protected areas, we allowed nature to take the helm?

If you think that goes without saying, just take a look in the filing cabinets of conservation areas and national parks. They are stuffed with maintenance and development plans that are far too eager to put sawmills, chain saws, and heavy machinery to work. In the end, such plans are neither aesthetically appealing nor ecologically beneficial, because these areas are created to save as many native species of trees as possible. We've already seen that most attempts at fixing things come to nothing, so why not simply trust that mechanisms that are millions of years old can still function without us?

AMONG ALL THE awful news about the worldwide destruction of forests, there are increasing notes of hope. More and more people want to protect existing forests and plant new ones. This desire to start over raises a question. Can we ever recreate these multifaceted ecosystems? The Brazilian rain forest offers some grounds for optimism. It is thought to be particularly vulnerable to changes wrought by civilization, because the soils on which it depends are so old. "Old" here is used in terms of geological eras during which these soils have barely changed at all. This is partly because since the Tertiary, which ended more than 2.6 million years ago, there has been no formation of new mountains, which means there has been barely any erosion or formation of new soil by weathering rocky slopes. This peaceful period is evident deep into the underground soil layers, and deep in this case means down to an impressive depth of 100 feet.

In most areas of the forest I manage, you can't dig down farther than about 2 feet before you hit an underground layer of gravel, and even the upper layers of soil contain a lot of small stones. In many tropical soils in the Amazon, in contrast, all the rock has been ground down into tiny particles. That might sound like nutrient-rich soil to you, but it's just the opposite. After having been rained on for hundreds of thousands of years, the soils have lost most of their nutrients—water seeping through the soil has washed them down far deeper than plant roots can reach.

The wide range of species we see today, as well as the exuberant growth of forests in these latitudes, seem to contradict this fact, but the fecundity of these forests is possible only because nutrients are held hostage in the rain forest ecosystem as an army of insects, fungi, and bacteria recycle everything that dies by eating and digesting and excreting it. Every trunk that rots, every leaf that gets eaten by insects and excreted as humus releases the minerals it contains, which are greedily taken up by roots and converted into living tissue once again. If you cut down the trees, this cycle of life is rudely interrupted.

Clearing by fire leaves a great deal of ash. Ash is simply nutrients in concentrated form, which are now exposed to heavy rains without any form of protection, which means they are flushed away in rivers, disappearing forever from the land. This is why the agriculture that follows such clearances is profitable for a short time only—that is to say, until the brief boost of fertility from the ash fizzles out. The ravaged soils that remain are barely capable of supporting new trees, and any trees that are planted struggle to survive—if they survive. True tropical diversity with millions of species depends on the return of all the fungi, insects, and vertebrates, all of which require such special conditions that their return is unlikely. Or is it?

Let's go back to recovery's ground zero. The forest is no more and the soils are exhausted. How can there ever be hope if nutrients have disappeared deep underground never to be seen again or have been washed by the rain into the nearest river? After all, there's no natural mechanism to pump them back to the surface or float them back from the distant ocean. Yet the situation is not hopeless, and the tortured land doesn't necessarily have to become a desert.

As far as minerals are concerned, the Sahara could be seen as a first responder. Dust storms blow from the desert into the air an enormous amount of tiny particles of dirt, which are then carried way up high on the wind from Africa to South America. There, the dusty cargo is washed down by regular heavy rain to fertilize the ground. Nearly 33 million tons arrive this way every year, including about 24,000 tons of phosphorus, which is a potent plant fertilizer.

Scientists from the Earth System Science Interdisciplinary Center (ESSIC) at the University of Maryland[2] analyzed seven years' worth of satellite images to estimate the amount of dust as accurately as they could. Estimates varied enormously, but they strongly suspected that the constant arrival of airborne fertilizer is offsetting the loss of nutrients washed away into the ground. That's the case, at least, for intact forests. When forests are felled, there is a marked increase in the rate at which minerals are lost. Damn. There doesn't seem to be any way out of this mess. Is the situation really hopeless? No, it's not, as we can see if we take a closer look at those clear-cuts in the Amazon. When large expanses of forest are cleared away, the remains of settlements appear. Human settlements.

A research team led by Jennifer Watling, who is now at the University of São Paulo, found 450 geoglyphs in the state of Acre, Brazil. Geoglyphs are earthworks laid out in geometric patterns, and in Acre, they consist of a series of trenches and berms distributed across 5,000 square miles. People must have cleared the forest to build them, but the sites reveal that the original inhabitants proceeded cautiously. The researchers found no evidence of large-scale clearing. Instead, they discovered a system of forest management spanning thousands of years. Wait a minute. Discovered? How is it possible to calculate the size of clearings made in the forest thousands of years ago?

The key here turns out to be microscopic particles of silica called phytoliths, which are found in some plants. These particles vary from plant to plant, which is helpful, but what is even more important is that unlike organic substances, which decompose quickly, these crystals last practically for ever. And so, it is possible to build up a picture of which plants were growing based on the frequency of different phytoliths.

Jennifer Watling and her team discovered that over the course of the four thousand years the Indigenous peoples were changing the forest in Acre, grass—a plant typically found in open spaces—never made up more than 20 percent of the vegetation, and the combination of trees was significantly altered. Around the human made constructions, the number of palm trees, which are important sources of both food and building materials, increased dramatically. Even today, more than six hundred years after the settlements were abandoned, a noticeable number of palm trees remains in areas close to the geoglyphs.

The researchers' findings are encouraging. First, we have a style of agroforestry—that is to say, a mixture of agriculture and silviculture in the same area—that clearly functioned for a very long time without having any great impact on the environment. What worked then should also work now, and it points to a way we could preserve as much forest as possible without excluding people. Second, after six hundred years, the forest has regenerated so well that before this discovery, scientists assumed this was virgin forest untouched by human hands. So, from now on, we should put more trust in forest ecosystems and no longer use the word "irretrievable" when describing them. And third, this is a message about climate that really makes me pay attention.

The Indigenous forest settlers carried out their system of land management over enormous areas, and as soon as the people disappeared, the forest everywhere recovered on a similarly large scale. The small areas devoted to agriculture were quickly overgrown by trees, the density of the forest increased, and a great deal of carbon was stored in the mighty trees. In fact, so much carbon was stored all at once that the research team thinks it's possible that this was what triggered the Little Ice Age—a period when temperatures dropped around the world—and not the erupting volcanoes I mentioned earlier.[3] From the fifteenth century into the nineteenth century, temperatures fell, and failed harvests and famine went hand in hand with cold, rainy summers and long, freezing winters. Was all of this caused by the recovery of the Amazonian rain forest?

Of course, no one wants to return to times of famine, but our problem today isn't cold but increasingly warm temperatures. The positive message from all this is that not only can we win back the original forests, but doing that could also steer the climate in the right direction. And to achieve this we don't even need to do anything. Just the opposite, in fact. We need to leave things alone—on as large a scale as possible.

Epilogue

I LOVE TO TELL stories. I also love to play the ukulele, even though it's been years now and I still can't play it very well. It's a bit different with storytelling, and that's because of feedback I've received from the public (perhaps even from you). I remember the first time I appeared on television, back in 1998. In those days, I offered a survival course deep in the forest, where participants had to survive the weekend equipped with nothing but a sleeping bag, a cup, and a knife. This was a dream topic for the media (under the headline "A Forester Who Eats Worms!"), and a camera team from the local television station came to the forest I manage to interview one of the groups—and me, of course.

I thought I acquitted myself rather well, and later I sat down proudly with my family to watch the segment on the local news. Instead of being impressed, they were soon pointing out the many "aahs" that punctuated every sentence.

"There's another one, Papa!" my children called out every couple of seconds with increasing delight. My delight, however, sank with every remark they made, and by the end of the program I was in a pretty bad mood. But, with the next interview, I was painfully careful to avoid saying "aah," and I gradually managed to garner a little praise for my appearances.

Something similar happened on the many guided tours I gave of the forest when I talked about subjects such as ecological forest management or took people through our forest cemetery, the Final Forest. No one corrected me or remarked on my verbal shortcomings, but there were always questions afterward. I soon noticed that I was using too many technical expressions and giving an impersonal, boring recitation of facts about something close to my heart: the wonderful forest ecosystem and the threat it was under. After I spoke, the reaction was subtle but still painful. As soon as those first eyelids fell, I knew I was being too dry. Over the years, an undertone of emotion crept in, which was much more in line with my personal thinking. In other words, I relaxed and let my heart do the talking instead of my brain.

Again and again, after the tours were over, participants asked where they could read more about the topics I'd explained. All I could do was shrug apologetically. At some point, my wife pressured me into writing at least a few pages so that we had something I could hand out to people who were interested in learning more. Back then, I had not the slightest interest in doing this. A friend suggested coming to one of my tours with a recording device and then writing a book. Hmm. I didn't like that idea much, either.

And so, on a vacation to Lapland, I sat down outside the camper armed with pad and pencil and began to capture on paper the ideas I talked about during my tours. I resolved that if after a year no publishing house had shown an interest in my book then I would check writing off my to-do list. I hadn't expected things to turn out the way they did. The small press Adatia (which is no longer in business) published my first book, *Wald ohne Hüter* (Forest without protectors), and I thought that would be that. But, as time went by, more books flowed from my pen, and I began to thoroughly enjoy myself.

Unfortunately, my work did not spark professional discussions among foresters about the way we treat forests. In hindsight, I can see that from the lobbyists' point of view, it's best not to enter into a public discussion of critical analyses. But recently, with my book *The Hidden Life of Trees*, criticism from professional forestry circles came to the fore, thanks to pressure from the many people who'd read the book and wanted to know why we needed to be going into forests with all that heavy machinery. However, instead of debating my ideas, most critics from the circles of forestry took a different approach.

The language I use is too emotional, they said. My descriptions make trees and animals seem human, and that is not scientifically correct. But can a language stripped of emotion even be called a human language? Don't we function mostly in accordance with our emotions? Are descriptions of nature only reliable when all processes are presented in biochemical terms and are dissected so precisely that you get the impression that plants and animals are fully automatic, genetically programmed biological machines? After all, it would be possible to describe all our own feelings and activities that way, yet that would in no way describe what's going on inside us and what enriches our lives. It is more important to me to state the facts so that people can understand them emotionally. And then I can lead them on a full sensory tour of nature, because that way I can communicate one thing above all: the joy our fellow creatures and their secrets can bring us.

Acknowledgments

THE NETWORK OF nature is too diverse to ever fit between the covers of a book, which means I needed to choose particularly impressive examples and connect them so that readers could see the big picture. My wife, Miriam, was a great help to me here. She read the manuscript many times with a critical eye. She didn't hesitate to point out passages that needed work, and she helped me see ways I could improve my explanations.

As always, my children, Carina and Tobias, were a source of inspiration. After interminable debates around the breakfast table or in front of the television (which devolved into a sort of electronic campfire), new ideas were sparked that demanded space in the book.

My colleagues Lidwina Hamacher and Kerstin Manheller carried the load at the Forest Academy in Hümmel. We were right in the middle of the time-intensive period of establishing the academy while I was working on the manuscript. Both of them understood when I was up against a deadline and needed to keep writing, and they simply took over my part in managing our undertaking.

The Mysteries of Nature trilogy—*The Hidden Life of Trees*; *The Inner Life of Animals*; and this book, *The Secret Wisdom of Nature*—would never have happened if the publisher hadn't believed from the outset that my message to the people who visited the forest I managed should reach a wider audience. My agent, Lars Schultze-Kossack, helped me through all the issues that cropped up along the way.

Heike Plauert of Ludwig Verlag, my publisher, made it easy for me, because she placed all her trust in me and let me get on with writing. That suited me very well, as I was working on different sections of the book at the same time—a process that takes some getting used to. My editor, Angelika Lieke, tactfully helped me polish the manuscript.

Beatrice Braken-Gülke in the publicity department managed the media so that I had room to breathe, though I would have been more than happy to answer all of their questions.

Last but not least, I would like to give a big thanks to Jane Billinghurst. I am very happy that her translations capture not only the meaning of the words I write but also the tone I wish to convey.

There are so many more people who are part of this process that unfortunately I cannot mention all of them. From the printer to the distributor to the booksellers, everyone has done their best to make sure that you get to hold this book in your hand. And I want to thank you most sincerely for choosing this book from the many good books out there so that you can make this journey through nature with me.

Notes

1/ Of Wolves, Bears, and Fish

1. MyYellowstonePark.com, "Wolf Reintroduction Changes Ecosystem," June 21, 2001, www.yellowstonepark.com/things-to-do/wolf-reintroduction-changes-ecosystem.
2. William J. Ripple, et al., "Trophic Cascades from Wolves to Grizzly Bears in Yellowstone," *Journal of Animal Ecology* 83, no. 1 (2014): 223–33, doi.org/10.1111/1365-2656.12123.
3. "Der Lübtheener Wolf wurde gezielt erschossen" (The Lübtheener wolf was shot on purpose), NABU (Naturbundschutz Deutschland), December 2016, www.nabu.de/news/2016/12/21719.html.
4. M. Holzapfel, et al., "Die Nahrungsökologie des Wolfes in Deutschland von 2001 bis 2012" (The food ecology of the wolf in Germany from 2001 to 2012), www.wolfsregion-lausitz.de/index.php/nahrungszusammensetzung.
5. "Wie viel Naturschutz verträgt unser Land?" (How much conservation can our country handle?) Statement by Olaf Tschimpke, president of NABU (Naturbundschutz Deutschland), on the television program *Hart aber fair* (Tough but fair), January 23, 2017, ARD (German public broadcasting), www.nabu.de/news/2017/01/21855.html.

6. A.D. Middleton, et al., "Grizzly Bear Predation Links the Loss of Native Trout to the Demography of Migratory Elk in Yellowstone," *Proceedings of the Royal Society B: Biological Sciences* 280, no. 1762 (May 15, 2013): 2013.0870, doi.org/10.1098/rspb.2013.0870.

2/ Salmon in the Trees

1. S.M. Gende, T.P. Quinn, et al., "Magnitude and Fate of Salmon-Derived Nutrients and Energy in a Coastal Stream Ecosystem," *Journal of Freshwater Ecology* 19, no. 1: 149, doi.org/10.1080/02705060.2004.9664522.

2. J. Robbins, "Why Trees Matter," *New York Times*, April 11, 2012, www.nytimes.com/2012/04/12/opinion/why-trees-matter.html.

3. T. Reimchen and M. Hocking, "Salmon-Derived Nitrogen in Terrestrial Invertebrates from Coniferous Forests of the Pacific Northwest," BMC *Ecology* 2 (March 19, 2002): 4, doi.org/10.1186/1472-6785-2-4.

4. C. Wolter, "Nicht mehr als dreimal in der Woche Lachs" (Salmon no more than three times a week), *Nationalpark-Jahrbuch Unteres Odertal* 4: 118–26, www.nationalpark-unteres-odertal.de/de/publikationen/nicht-mehr-als-dreimal-der-woche-lachs.

5. "Quatsch anfangen" (Talking nonsense), *Der Spiegel*, vol. 38 (1988).

6. ARGE Ahr, "Bejagung des Kormorans" (Hunting cormorants), October 25, 2016, www.arge-ahr.de/tag/kormoran.

7. Anne Helmenstine, "Elements in the Human Body and What They Do," *Science Notes*, May 20, 2015, sciencenotes.org/elements-in-the-human-body-and-what-they-do/.

8. A. Oita, et al., "Substantial Nitrogen Pollution Embedded in International Trade," *Nature Geoscience* 9 (January 25, 2016): 111–15, doi.org/10.1038/NGEO2635.

9. Anne Post, "Why Fish Need Trees and Trees Need Fish," *Alaska Fish & Wildlife News*, November 2008, www.adfg.alaska.gov/index.cfm?adfg=wildlifenews.view_article&articles_id=407.

3/ Creatures in Your Coffee

1. Axel Bojanowski, "Forscher rätseln über seltsame Tiefenwesen" (Researchers puzzle over strange life-forms), *Der Spiegel* online, December 13, 2013, www.spiegel.de/wissenschaft/natur/mikroben-ursprung-des-lebens-kilometer-unter-erde-moeglich-a-938358-druck.html.

2. Bundesministerium für Umwelt, Naturschutz und Reaktorsicherheit (BMU), *Grundwasser in Deutschland* (Groundwater in Germany), Referat Öffentlichkeitsarbeit, Berlin, August 2008, 7.

3. Gianfranco Novarino, et al., "Protistan Communities in Aquifers: A Review," *FEMS Microbiology Reviews* 20 (1997): 261–75, doi.org/10.1111/j.1574-6976.1997.tb00313.x, onlinelibrary.wiley.com/doi/10.1111/j.1574-6976.1997.tb00313.x/full.

4. R. Sender, S. Fuchs, R. Milo, "Revised Estimates for the Number of Human and Bacteria Cells in the Body," *PLOS Biol* 14, no. 8 (2016): e1002533, doi.org/10.1371/journal.pbio.1002533.

4/ Why Deer Taste Bad to Trees

1. B. Ochse, et al., "Salivary Cues: Simulated Roe Deer Browsing Induces Systemic Changes in Phytohormones and Defence Chemistry in Wild-Grown Maple and Beech Saplings," *Functional Ecology* 31, no. 2 (August 8, 2016): 340–49, doi.org/10.1111/1365-2435.12717.

5/ Ants—Secret Sovereigns

1. Christoph Drösser, "Ein Haufen Ameisen" (A whole heap of ants), *Die Zeit* online, March 20, 2008, www.zeit.de/2008/13/Stimmts-Ameisen-und-Menschen.

2. W. Jirikowski, "Wichtige Helfer im Wald: hügelbauende Ameisen" (Important forest helpers: hill-building ants), *Der Fortschrittliche Landwirt*, Graz (14): 105–7.

3. T. Oliver, et al., "Ant Semiochemicals Limit Apterous Aphid Dispersal," *Proceedings of the Royal Society B* 274, no. 1629 (December 22, 2007): 3127–32, doi.org/10.1098/rspb.2007.1251.

4. T. Mahdi and J.B. Whittaker, "Do Birch Trees (*Betula pendula*) Grow Better if Foraged by Wood Ants?" *Journal of Animal Ecology* 62, no. 1 (January 1993): 101–16, doi.org/10.2307/5486.

5. J.B. Whittaker, "Effects of Ants on Temperate Woodland Trees," in C.R. Huxley and D.F. Cutler, eds., *Ant-Plant Interactions* (New York: Oxford University Press, 1991), 67–79.

6/ Is the Bark Beetle All Bad?

1. Government of British Columbia, "Mountain Pine Beetle Projections," www2.gov.bc.ca/gov/content/industry/forestry/managing-our-forest-resources/forest-health/forest-pests/bark-beetles/mountain-pine-beetle/mpb-projections.

2. H. Rosner, "The Bug That's Eating the Woods," *National Geographic*, April 2015, ngm.nationalgeographic.com/2015/04/pine-beetles/rosner-text.

7/ The Funeral Feast

1. X. Gu and R. Krawczynski, "Tote Weidetiere—staatlich verhinderte Forderung der Biodiversität" (Dead grazing animals—federally impeded requirements for biodiversity), *Artenschutzreport*, no. 28 (2012): 60–64.

2. Ibid.

3. "Die Rückkehr des Knochenfressers" (The return of the bone eater), spectrum.de, September 24, 2010, www.spektrum.de/news/die-rueckkehr-des-knochenfressers/1046860.

4. Club300.de, "Rarities Germany: Gänsegeier" (German rarities: griffon vultures), www.club300.de/alerts/index2.php?id=203.

5. C. Westerhaus, "Weibchen lassen Männchen während der Brutpflege abblitzen" (Females rebuff males when caring for their young) Deutschlandfunk, March 23, 2016, www.deutschlandfunk.de/totengraeber-kaefer-weibchen-lassen-maennchen-waehrend-der.676.de.html?dram:article_id=349257.

8/ Bring Up the Lights!

1. H.U. Schnitzler and O.W. Henson, "Sonar Systems in Microchiroptera," in R.G. Busnel, et al., *Animal Sonar Systems* (New York. Plenum Press, 1980).

2. Hannah M. Moir, et al., "Extremely High Frequency Sensitivity in a 'Simple' Ear," *Biology Letters* 9, no. 4 (May 8, 2013), doi.org/10.1098/rsbl.2013.0241.

3. "Wo tanzt das Glühwürmchen?" (Where does the glowworm dance?), www.laternentanz.eu/Content/Informations/Living.aspx.

4. David J. Merritt and Sakiko Aotani, "Circadian Regulation of Bioluminescence in the Prey-Luring Glowworm, *Arachnocampa flava*," *Journal of Biological Rhythms* 23, no. 4 (August 2008): 319–29, doi.org/10.1177/0748730408320263.

5. Wynne Parry, "Fireflies' Unique Flashes Help Distinguish Species," LiveScience, March 29, 2012, www.livescience.com/19376-firefly-glow-signals.html.

6. T. Eisner, et al., "Firefly 'Femmes Fatales' Acquire Defensive Steroids (Lucibufagins) from Their Firefly Prey," *PNAS* 94, no. 18 (September 2, 1997): 9723–28, doi.org/10.1073/pnas.94.18.9723.

9/ Sabotaging the Production of Iberian Ham

1. Michael Stang, "Die eigenwilligen Flugrouten der Kraniche" (The self-selected migration routes of cranes), www.deutschlandfunk.de/globales-kommunikationsnetz-bei-zugvoeglen-die.676.de.html?dram:article_id=321788.

2. Gregor Rolshausen, et al., "Contemporary Evolution of Reproductive Isolation and Phenotypic Divergence in Sympatry along a Migratory Divide," *Current Biology* 19, no. 24 (December 3, 2009): 2097–101, doi.org/10.1016/j.cub.2009.10.061.

3. Katy Sewall, "The Girl Who Gets Gifts from Birds," *BBC News Magazine* online, www.bbc.com/news/magazine-31604026.

10/ How Earthworms Control Wild Boar

1. W. Arnold, et al., "Nocturnal Hypometabolism as an Overwintering Strategy of Red Deer (*Cervus elaphus*)," *American Journal of Physiology, Regulatory, Integrative, and Comparative Physiology* 286, no. 1 (January 1, 2004): R174–R181, doi.org/10.1152/ajpregu.00593.2002.

2. Blick Acktuell, "Hohe Rotwilddichte im Kesselinger Tal wird zu Problem" (The high density of red deer in the Kesslinger Valley is becoming a problem), www.blick-aktuell.de/Bad-Neuenahr/Hohe-Rotwilddickte-im-Kesselinger-Tal-wird-zu-Problem-27341.html.

3. U. Dohle, "Besser: Wie mästet Deutschland?" (Better: How does Germany fatten up?), *Ökojagd*, February 2009, 14–15.

4. Forstbotanische Garten (forest-botanical garden), Georg-August-Universität-Göttingen, "Stieleiche" (The common oak), www.uni-goettingen.de/de/blüten-samen-und-früchte/16692.html.

5. N. Hahn, "Raumnutzung und Ernährung von Schwarzwild" (Territorial use and feeding of wild boar), *LWF aktuell* 35, 32–34, www.waldwissen.net/wald/wild/management/lwf_raum_schwarzwild/index_DE.

6. Axel Weiss, "Sauenmast im Westerwald" (Feeding sows in Westerwald), www.swr.de/blog/umweltblog/2008/10/18/sauenmast-im-westerwald/.

7. Karl-Maria ImBoden, "Regenwurm" (Earthworm), www.regenwurm.ch/de/leistungen.html.

8. S. Blome and M. Beer, "Afrikanische Schweinepest" (African swine pest), *Berichte aus der Forschung, FoRep* 2/2013, Friederich-Löffler-Institut, Insel Reims.

11/ Fairy Tales, Myths, and Species Diversity

1. Bundesamt für Naturschutz (BfN) (Federal Agency for Nature Conservation), "Artenschutz-Report 2015, Tiere und Pflanze in Deutschland" (Conservation report 2015, Animals and plants in Germany), Bonn, May 2015, 12.

2. Federal Ministry of Food and Agriculture, "The Forests in Germany: Selected Results of the Third National Forest Inventory," October 2014, translated January 2015, www.bmel.de/SharedDocs/Downloads/EN/Publications/ForestsInGermany-BWI.pdf?__blob=publicationFile.

3. "Neue Tierart entdeckt" (New species discovered), Pressemitteilung des Helmholtz-Zentrums für Umweltforschung UFZ (Press release from the Helmholtz Center for Environmental Research—UFZ), March 20, 2015, www.ufz.de/index.php?de=35747.

4. E. Dressaire, et al., "Mushroom Spore Dispersal by Convectively Driven Winds," Cornell University Library, December 2015, arXiv 1512.07611v1 [physics.bio-ph].

5. C. Pietschmann, "Pilzgespinst in Wurzelwerk" (Fungal threads in root systems), Max-Planck-Institut für molekulare Pflanzenphysiologie (Max Planck Institute for Molecular Plant Physiology), December 21, 2011, www.mpimp-golm.mpg.de/5630/news_publications_4741538.

6. Anne Casselman, "Strange but True: The Largest Organism on Earth Is a Fungus," *Scientific American*, October 4, 2007, www.scientificamerican.com/article/strange-but-true-largest-organism-is-fungus/.

7. G. Möller, "Struktur- und Substratbindung holzbewohnender Insekten, Schwerpunkt Coleoptera-Käfer" (Structure and substrate-

specific binding of insects that live in wood, with a focus on Coleoptera), Dissertation zur Erlangung des akademischen Grades des Doktors der Naturwissenschaften (Dr.rer.nat.), eingereicht im Fachbereich der Biologie, Chemie, Pharmazie der Freien Universität Berlin (Ph.D. dissertation in natural sciences, Faculty of Biology, Chemistry, Pharmacy, Free University Berlin), March 2009, 35–36.

12/ What's Climate Got to Do with It?

1. K. Naudts, et al., "Europe's Forest Management Did Not Mitigate Climate Warming," *Science* 351, no. 6273 (February 5, 2016): 597–600, doi.org/10.1126/science.aad7270.

2. Michelle Hampson, "Centuries of European Forest Management Have Not Cooled Climate," American Association for the Advancement of Science, February 3, 2016, www.aaas.org/news/science-centuries-european-forest-management-have-not-cooled-climate.

3. Jason Kirby, et al., "Ion-Induced Nucleation of Pure Biogenic Particles," *Nature* 533 (May 26, 2016): 521–26, doi.org/10.1038/nature17953.

4. R. Wengenmayr, "Staub, an dem Wolken wachsen" (Dust that grows clouds), Mitteilung der Max-Planck-Gesellschaft, February 22, 2010, Max Planck Institute for Chemistry, Mainz.

5. M. Dobbertin and A. Giuggiolo, "Baumwachstume und erhöhte Temperaturen" (Tree growth and rising temperatures), *Forum für Wissen* 206: 35–45.

6. Institute of Veterinary Public Health, University of Veterinary Medicine Vienna, World Maps of Köppen-Geiger Climate Classification, koeppen-geiger.vu-wien.ac.at/.

7. Karl Gartner and Markus Neumann, "Wie schützen sich Waldbäume vor extremer Kälte?" (How do forest trees protect themselves against extreme cold?), waldwissen.net, March 6, 2012, www.waldwissen.net/wald/klima/wandel_co2/bfw_schrumpfen_baumstamm/index_DE.

8. G.H. Miller, et al., "Abrupt Onset of the Little Ice Age Triggered by Volcanism and Sustained by Sea-Ice/Ocean Feedbacks," *Geophysical Research Letters* 39, no. 2 (January 2012): L02708, doi.org/10.1029/2011GL050168.

13/ It Doesn't Get Any Hotter Than This

1. D. Kraus, F. Krumm, and A. Held, "Waldbrände schaffen Artenvielfalt" (Fire creates species diversity), www.waldwissen.net/waldwirtschaft/schaden/brand/fva_waldbrand_artenvielfalt/index_DE.

2. F. Berna, et al., "Microstratigraphic Evidence of In Situ Fire in the Acheulean Strata of Wonderwerk Cave, Northern Cape Province, South Africa," PNAS 109, no. 20: E1215–E1220, doi.org/10.1073/pnas.1117620109.

3. P. Bethge, "Ich koche, also bin ich" (I cook, therefore I am), *Der Spiegel* 52, 2007, 126–29.

4. Peter Hirschberger, *Wälder in Flammen: Ursachen und Folgen der weltweiten Waldbrände* (Forests in flame), WWF Deutschland, Berlin, July 2011, 33.

14/ Our Role in Nature

1. Monash University, News and Events, "Humans Caused Australia's Megafaunal Extinction," January 20, 2017, www.monash.edu/news/articles/humans-caused-australias-megafaunal-extinction.

2. Thomas Litt, "Waldland Mitteleuropa: Die Megaherbivorentheorie aus paläobotanischer Sicht" (Central Europe as forest: The megaherbivore theory from a paleobotanical perspective), in Tagungsband "Grosstiere als Landschaftsgestalter—Wunsch oder Wirklichkeit?" (Conference transcript: Large animals as shapers of the landscape—fact or fiction?) Freising, August 2000, Bayerisches Staatsministerium für Ernährung, Landwirtschaft und Forsten (Bavarian Ministry of Food, Agriculture and Forestry), no. 27: 50–57, www.gbv.de/dms/tib-ub-hannover/320198219.pdf.

3. Hubert Weiger, "Chancen und Risiken der Megaherbivorentheorie" (Prospects and risks of the megaherbivore theory), in Tagungsband "Grosstiere als Landschaftsgestalter—Wunsch oder Wirklichkeit?" (Conference transcript: Large animals as shapers of the landscape—fact or fiction?) Freising, August 2000, Bayerisches Staatsministerium für Ernährung, Landwirtschaft und Forsten (Bavarian Ministry of Food, Agriculture and Forestry), no. 27: 3–5, www.gbv.de/dms/tib-ub-hannover/320198219.pdf.

4. BUND, Friends of the Earth Germany, "Ein Rettungsnetz für die Wildkatze" (A safety net for wild cats), www.bund.net/themen_und_projekte/rettungsnetz_wildkatze/.

5. J.R. Moore, et al., "Anthropogenic Sources Stimulate Resonance of a Natural Rock Bridge," *Geophysical Research Letters* 43, no. 18 (September 28, 2016): 9669–76, doi.org/10.1002/2016GL070088.

15/ The Stranger in Our Genes

1. F. Ramirez Rozzi, et al., "Cutmarked Human Remains Bearing Neandertal Features and Modern Human Remains Associated with the Aurignacian at Les Rois," *Journal of Anthropological Sciences* 87 (2009): 153–85.

2. C. Simonti, et al., "The Phenotypic Legacy of Admixture between Modern Humans and Neanderthals," *Science* 351, no. 6274 (February 2, 2016): 737–41, doi.org/10.1126/science.aad2149.

3. Ibid.

4. M. Kuhlwilm, et al., "Ancient Gene Flow from Early Modern Humans into Eastern Neanderthals," *Nature* 530 (February 25, 2016): 429–33, doi.org/10.1038/nature16544.

5. G. Arora, et al., "Did Natural Selection for Increased Cognitive Ability in Humans Lead to an Elevated Risk of Cancer?" *Medical Hypotheses* 73, no. 3 (September 2009): 453–56.

16/ The Old Clock

1. Example of how coppicing is portrayed in the promotional literature of a federal forest commission: Landesforsten Rheinland-Pfalz (State Forests of Rhineland-Palatinate), "Niederwald in Rheinhessen—ein neues Projekt" (Coppicing in Rheinhessen—a new project), www.wald-rlp.de/en/forstamt-rheinhessen/wald/niederwaldprojekt.
2. H. Yu, et al., "The Fertilizing Role of African Dust in the Amazon Rainforest: A First Multiyear Assessment Based on Data from Cloud-Aerosol Lidar and Infrared Pathfinder Satellite Observations," *Geophysical Research Letters* 42 (March 18, 2015): 1984–91, doi.org/10.1002/2015GL063040.
3. J. Watling, et al., "Impact of Pre-Columbian Geoglyph Builders on Amazonian Forests," PNAS 114, no. 8 (2017): 1868–73, doi.org/10.1073/pnas.1614359114.

Index

THE DAVID SUZUKI INSTITUTE is a non-profit organization founded in 2010 to stimulate debate and action on environmental issues. The Institute and the David Suzuki Foundation both work to advance awareness of environmental issues important to all Canadians.

We invite you to support the activities of the Institute. For more information please contact us at:

David Suzuki Institute
219 – 2211 West 4th Avenue
Vancouver, BC, Canada V6K 4S2
info@davidsuzukiinstitute.org
604-742-2899
www.davidsuzukiinstitute.org

Cheques can be made payable to The David Suzuki Institute.